Springer Theses

Recognizing Outstanding Ph.D. Research

Aims and Scope

The series "Springer Theses" brings together a selection of the very best Ph.D. theses from around the world and across the physical sciences. Each volume has been selected for its scientific excellence by an internationally renowned institute or department. For greater accessibility to non-specialists, the published versions include an extended introduction, as well as a foreword by the student's supervisor explaining the special relevance of the work for the field. The series as a whole provides a valuable resource both for newcomers to the research fields, and for other scientists seeking detailed background information on special questions. Finally, it serves as an accredited documentation of the valuable contributions made by today's younger generation of scientists.

How to Apply

Theses prepared at internationally leading universities and institutes may be nominated for publication by supervisors with the endorsement of the head of department. Each such department is entitled to nominate one thesis per year. Enquiries and requests for a nomination form should be sent, together with the thesis, to Ms. Nathalie Berg (nathalie.berg@springer.com).

Eligibility Criteria

To qualify for consideration, theses should fulfil the following criteria:

- They must be written in good English.
- The topic should fall within the confines of Chemistry, Physics, Earth Sciences, Engineering and related interdisciplinary fields such as Materials, Nanoscience, Chemical Engineering, Complex Systems and Biophysics.
- The work reported in the thesis must represent a significant scientific advance.
- A maximum of 30% of the text may be a verbatim reproduction from the author's previous publications.
- They must have been examined and passed during the 12 months prior to nomination.
- The theses should have a clearly defined structure including an introduction accessible to new PhD students and scientists not expert in the relevant field.
- After acceptance, theses should be supplemented by a "Supervisor's Foreword" outlining the wider scientific relevance of the work

Indexed by zbMATH.

Rourav Basak

Nucleation, Fabrication, and Competition in Antiferromagnetic Textures Studied with Coherent Resonant Soft X-ray Scattering

Doctoral thesis accepted by
University of California, San Diego, La Jolla, USA

Springer

Author
Dr. Rourav Basak
University of California, San Diego
La Jolla, CA, USA

Supervisor
Prof. Alex Frañó
University of California, San Diego
La Jolla, CA, USA

ISSN 2190-5053 ISSN 2190-5061 (electronic)
Springer Theses
ISBN 978-3-032-19198-4 ISBN 978-3-032-19199-1 (eBook)
https://doi.org/10.1007/978-3-032-19199-1

This Springer imprint is published by the registered company Springer Nature Switzerland AG
The registered company address is: Gewerbestrasse 11, 6330 Cham, Switzerland

To Ma, Baba, and Esha.

Supervisor's Foreword

The study of antiferromagnetic materials has become a central pursuit in modern condensed matter physics, because antiferromagnetic interactions sit at the heart of many quantum mechanical phenomena and appear throughout the phase diagrams of the most intriguing correlated materials. They tend to be present in the proximity of remarkable emergent behaviors, often signaling regions in which the electronic system is finely balanced and sensitive to competing influences. As a result, these materials often harbor subtle forms of spin order, intricate phase competition, and real-space textures that reflect the deeper structure of strongly interacting electrons. To access these features, we rely on experimental tools capable of resolving essential spatial and momentum information, and coherent scattering has become one of the most incisive approaches for this purpose. Coherent X-ray methods allow us to follow the nucleation, evolution, and reorganization of antiferromagnetic domains, offering insights that are not accessible through conventional probes. The knowledge gained through these measurements advances our fundamental understanding of correlated matter, which is the foundation upon which future technological implementations of this will inevitably rest.

Within this scientific landscape, Dr. Rourav Basak has created a dissertation that is remarkable in both its clarity and its ambition. The work begins with two chapters that are remarkable in their clarity, structure, and pedagogical intention. Chapters 1 and 2 provide a carefully organized introduction to the physics of correlated electron systems and to the principles underlying coherent resonant X-ray scattering. The discussion of antiferromagnetic textures, pattern formation, coherence, interference, and resonant scattering is presented with a level of clarity and precision that makes these foundational ideas easily accessible to readers entering the field. The accompanying figures are crafted with care, and the progression from conceptual background to experimental technique is seamlessly constructed. These opening chapters establish the scientific motivation and the methodological framework that support the original results developed in the later parts of the thesis.

The core research of the dissertation is presented through a series of studies that explore antiferromagnetic textures from complementary perspectives. The first original project focuses on the artificial creation and control of antiferromagnetic patterns in rare-earth nickelate thin films. Here, Rourav and his collaborators used focused helium-ion implantation as a local structural tuning mechanism, modifying the balance among intertwined charge, lattice, and spin degrees of freedom in $NdNiO_3$ and $PrNiO_3$. The work demonstrates how bond-order–driven charge disproportionation and antiferromagnetic order can be manipulated to generate spatially defined magnetic textures. Confirmed using resonant photoemission microscopy and resonant coherent scattering, the chapter establishes a versatile and conceptually elegant approach for engineering correlated-electron landscapes.

Chapter 4 examines the phenomenon of phase competition in Fe-intercalated NbS_2, a material where antiferromagnetic order and charge modulations intersect in subtle and surprising ways. The discovery of a previously unreported charge-ordered state and the careful elucidation of its coupling to the antiferromagnetic order parameter are remarkable contributions of Rourav's dissertation. Then, coherent scattering is used to track the evolution of stripe-like textures, revealing their melting and disordering behavior across temperature or via the competition of another magnetic ground state. The analysis ties together reciprocal- and real-space perspectives and offers a refined picture of how competing phases negotiate the mesoscale spatial landscape of a correlated system. These highly original results contribute meaningfully to ongoing discussions about phase competition in quantum materials.

The final chapter introduces a conceptually distinct and forward-looking development: a multiple–beam-spot coherent scattering geometry. By generating several coherent illumination points separated in a controlled manner, the resulting diffraction pattern acquires new forms of spatial encoding that allow correlations to be disentangled along different directions. This approach expands the information content available in coherent scattering experiments and reflects the kind of imaginative, boundary-expanding thinking that has characterized Rourav's work throughout his time in my group. It provides a promising pathway for future studies of complex antiferromagnetic and charge-ordered textures.

Beyond the scientific results themselves, this dissertation also reflects the qualities that have defined Rourav's presence in the laboratory and within multi-institutional collaborations. He brings a together a wonderful combination of creativity, persistence, and genuine enthusiasm for the scientific process. His work often sits at the intersection of several communities, and he approaches these collaborations with openness and intellectual curiosity. Equally meaningful is the spirit with which he pursued his science: he brought generosity, encouragement, and a sense of shared purpose to every interaction. He did this all throughout his endeavor, from those long nights at beamlines to those engaging brainstorming discussions about the physical interpretation of experimental results and their meaning. These human qualities are not always visible on the page, but they were felt clearly by everyone who worked alongside him. They strengthened every project represented in this thesis: the experiments were richer, the interpretations sharper, and the scientific conversations more energizing because of his thoughtful and joyful participation.

Taken together, this thesis presents a coherent, original, and beautifully executed exploration of antiferromagnetic textures and coherent scattering methodologies. It stands as a record of Rourav's remarkable journey. It brings together years of careful measurement, conceptual development, and inventive thinking, presented with clarity and genuine pedagogical care. I am confident that the contributions made here will influence future efforts in coherent scattering and correlated electron physics. It is a joy to introduce this work to the wider community, and it has been a privilege to accompany Rourav Basak during this formative stage of his scientific career.

La Jolla, USA
February 2026

Alex Frañó

Parts of this thesis have been published in the following journal articles:

- *Bluschke, M., *Basak, R., Barbour, A., Warner, A.N., Fürsich, K., Wilkins, S., Roy, S., Lee, J., Christiani, G., Logvenov, G. and Minola, M., 2022. **Imaging mesoscopic antiferromagnetic spin textures in the dilute limit from single-geometry resonant coherent x-ray diffraction**. Science Advances, 8(29), p.eabn6882.
- *Wu, S., *Basak, R., Li, W., Kim, J.W., Ryan, P.J., Lu, D., Hashimoto, M., Nelson, C., Acevedo-Esteves, R., Haley, S.C. and Analytis, J.G., 2023. **Discovery of charge order in the transition metal dichalcogenide Fe_xNbS_2**. Physical Review Letters, 131(18), p.186701.

(* indicates equal contribution)

Acknowledgements I would like to thank my advisor, Prof. Alex Frañó, for his unwavering support and guidance. He inspired me to explore a vast variety of problems which gave me the opportunity to diversify my horizon. I am deeply indebted to him. His teaching goes beyond the pages into the world of abstractness, while keeping the foot grounded in the structure, like music. I would also like to thank my committee members Prof. Daniel Arovas, Prof. Eric Fullerton, Prof. Tenio Popmintchev, and Prof. Oleg Shpyrko for their guidance, teaching, and support during my Ph.D. at UCSD. Among many other teachings, the discussions about coherent X-ray scattering with Prof. Oleg Shpyrko and Prof. Sunil Sinha were particularly instrumental in this dissertation.

I am grateful to the postdoctoral researchers who helped me understand the nuances of the fields. Dr. Martin Blushcke and Dr. Ivan Zaluzhnyy taught me coherent scattering patiently. I learned about strongly correlated systems from Dr. Alejandro Ruiz and Dr. Sarmistha Das. Dr. ShanWu taught me neutron scattering and strongly correlated magnetic systems. Their mentoring, support, and encouragement were very precious along my journey.

Beamtimes at sychrtron facilities around the world were a significant portion of my research. I realized beamtimes are a great to develop friendships. Boyan, Erik, Brandon, Nelson, Elliot, Biswajit, Mayia, Dong, Francis, Robin, Nicolo, Josh, Holden, Andy, Nathan, Shan, Youzhe, Sarmistha, Alejandro, Ivan, Kirsty, and Wenxin: thank you all for the fun, for the shared journey towards success and failure, and the stimulating discussions along the way. The hardship of the beamtimes became a cakewalk with them. I am also extremely grateful to share the 3rd floor of Mayer hall with them and Reno, Kate, Amy, Ash, Ashley, Adolfo, and Quinn.

I loved the physics classes in UCSD and somehow I learned every course in Physics from a condensed matter physics professor. Thanks to Prof. Dan Arovas, Prof. Nigel Goldenfeld, Prof. Yi-Zhuang You, Prof. John McGreevy, Prof. Michael Fogler, and Prof. Tarun Grover for the fantastic classes. I learned a lot and enjoyed them thoroughly. My professors in New York University: Prof. L. Andrew Wray, Prof. Lorcan Forlan, Prof. Vladimir Tsifrinovich, Prof. Edward Wolf, and Prof. David Grier—I am thankful for the classes, research opportunities, and guidance. Apart from classes, I learned significantly from the beamline scientists who are a library of knowledge by themselves. Dr. Fanny Rodolakis, Dr. Andi Barbour, Dr. Yi-de Chuang, and Dr. Eugene Weshcke were instrumental in shaping me as an experimenter. Among other beamline scientists, I was fortunate enough to work with Dr. Claudio Mazzoli, Dr. Xiao Chen, Dr. Andrei Fluerasu, Dr. Christie Nelson, Dr. Raul Acevedo-Esteves, Dr. Rahul Jungid, Dr. Jonathan Pelliciari, Dr. John Freeland, Dr. Jessica McChesney, Dr. Philip Ryan, Dr. Jong-Woo Kim, Dr. Eric Dufresne, Dr. Suresh Narayan, Dr. Christoph Klewe, Dr. Alpha N'Diye, Dr. Padraic Shafer, Dr. Sujoy Roy, Dr. Sophie Morley, Dr. Andreas Scholl, Dr. Anoitne Wojdyla, and Dr. Vivek Thampy. I want to thank them all for their dedication, shared enthusiasm, teaching, and support during the beamtimes and beyond.

I am also thankful to Dr. Uday Goteti, Dr. Frances Allen, Dr. Nirjhar Sarkar, and Prof. Robert Dynes for the samples related to the focused helium implantation. Tianxing, Nareg, Junjie, Ben, Henry, Ali, and Prof. Ivan K. Schuller were also earnestly helpful in characterizing the resistance of the hundreds of thin films used for this research. For the samples related to Fe intercalated NbS_2, I am grateful to Dr. Shanon Haley and Prof. James Analytis. I thoroughly enjoyed working on this material with my Berkeley collaborators Dr. Shan Wu and Prof. Robert Birgeneau. Frequent meetings with them definitely shaped me as a better condensed matter physicist. I thank Dr. Martin Sundermann and Dr. Hasan Yavas for writing the orbital imaging paper which gave me sheer joy from such a beautiful and nuanced experiment. Thanks to my larger X-ray scattering community for bringing me up: Prof. Yu He, Prof. Eduardo di Silva Nato, Prof. Ming Yi, Prof. Riccardo Comin, Dr. Abe Levitan, Ahmet Demir, Dayne Sasaki, Dr. Andrei Singer, Dr. Suchismita Sarkar, Dr. Nathan Giles-Donovan, Dr. Youzhe Chen, Dr. Jissop Oh, Kirsty Scott, Tyler Warner, Winxin Li. If it takes a village to raise a Ph.D. student, you created the one for me. Thanks to Natalie Lam, Josh Evans, and Holden Bauer with whom I embarked on crazy experiments during their summer holidays. I am deeply honored to have shared the journeys with you. I am also thankful for the support system of UCSD which manifested through Sharmila Poddar, Catherine McConney, Brad Hanson, Feifei Yu, Tina Silva, Robin Knoxx, Maio Rojas, Ray Descoteaux, and Lester Brooks.

My friends in San Diego: Ieva and Boyan, Rinki and Abhra, Varun, Sridip, Ahanjit, Shouvik, Nirjhar, Evgueni, Sourjya and Soumi, my friends in New York: Shayantani, Amitabha, Mainak, Erica, Igor, Paul, Alex—thank you for being the support system through thick and thin, through good food and drinks. Arkarup, Sanchari, Sanghita, Shyamal, and Joydip Ghosh thank you for guiding me during the early days of Physics and graduate applications. Thank you Prof. Raj Gandhi for meeting me and giving me directions on how to find my path.

I am forever indebted to my friend Pratim. I am sure this journey would not have been possible without your crucial support when I needed them the most. Since my father, Ranendra Nath, taught me the first lesson in geometry, I was hooked to Physics. Thank you for teaching me so nicely during my childhood. My mother, Krishna, taught me how to question and my father taught me how to be curious. My parents provided the essential environment and taught me everything that I needed to learn. Thank you Baba—Ma for being the amazing people you are. I found a similar set of amazing people as my parents-in-laws, Subrata and Nandita. Thank you for the confidence and support you provide every day. Finally, Esha, my wife—I don't even know where to begin. I am so grateful to have met you and shared last 15 years with you. Your love, support, ideas, and above all sacrifices were vital in this journey.

Competing Interests The author has no competing interests to declare that are relevant to the content of this manuscript.

Contents

Chapter 1
Pattern Formation in Antiferromagnets

Abstract Pattern formation in strongly correlated electron systems offers both fundamental scientific insight and potential technological impact, particularly for information processing. In this chapter, we introduce the concept of pattern formation within the context of strongly correlated electron physics, with an emphasis on antiferromagnets. We then highlight the technological relevance of antiferromagnetic textures as functional building blocks. A broad overview of experimental approaches to probe such textures is provided, setting the stage for the discussions that follow. Finally, we outline how the subsequent chapters build on these foundations to explore pattern formation in correlated systems in greater depth.

1.1 Introduction to Pattern Formation

A homogeneous universe would have been a boring universe. Our captivating universe showcases intricate structures at different length scales. A simple short-ranged interaction between nearest neighbors can manifest long range orders of thermodynamically large number of constituents and spontaneously break the symmetry. The system is considered to have gone through a 'phase transition'. The collective behavior of the constituents in a long range order can behave significantly different from the fundamental property of the constituents [1]. One striking example is superconductivity, where individual electrons, instead of repelling each other, forms 'cooper pair' to conduct energy without dissipation. That knowing the physics of constituent particles does not automatically allow us to predict the properties of the system at a larger length scale where collective behavior becomes important, is the central theme of condensed matter physics—one of the many things that makes our lives significantly interesting. It is during the phase transitions we encounter the mesmerizing pattern formations. We first discuss the length scale where the material under investigation appears inhomogeneous.

R. Basak, *Nucleation, Fabrication, and Competition in Antiferromagnetic Textures Studied with Coherent Resonant Soft X-ray Scattering*, Springer Theses,
https://doi.org/10.1007/978-3-032-19199-1_1

Length-Scales of Interest

Correlation length (ξ) of a system under phase transition defines the characteristic length scale over which the fluctuations are correlated. Near a critical point in a phase diagram, this characteristic length diverges. Away from the critical point this length becomes finite. Furthermore, as we lower the temperature, going away from the critical point, the order parameter becomes stronger, facilitating a region of ordered constituents within which the fluctuations are low, known as domains. Thus, away from the critical point, the separation of the domain boundary can be considered equivalent to the correlation length—this is the typical length scale in an ordered system which hosts homogeneity. Thus, we see that a system under phase transition typically offers us the opportunity to study an inhomogeneous material system without even considering the effect of defects and distortion. In our research, we focus primarily on such systems.

The length scale of our problem begins at $\mathring{A}$ where the ions are organized precisely to form a crystal structure. The typically discussed interactions in condensed matter physics range from nearest neighbor interaction to a few nanometers. Furthermore, we consider a thermodynamically large number of ions and electrons as the constituents. Thus, statistical properties of this ensemble dictate the likely state of the system. Furthermore, at this length scale of nearest neighbor interaction, quantum effects also become important. Different degrees of freedom of electrons follow the symmetry and topology of the space they span. In strongly correlated electron systems, such as high temperature superconductors, different degrees of freedom couple on top of the many body interaction they exhibit. Thus, such strongly interacting many body quantum systems become quite complex to understand. The domain sizes in these complex system ranges from a few nm to a few tens of nm. Competition of domains of two different ground states such as charge density waves and high temperature superconductivity, or recent experimental advances in creating lateral Moiré heterostructures in bi-layer graphene [2] inspired us to study pattern formation in quantum materials.

Motivation

One central motivation of this thesis is that the ability to manipulate and study meso-scale domain textures which offers us to closely monitor the effect of quantum degrees of freedom. The complex inhomogeneous texture of these materials was argued to be correlated with the observation of novel electronic phases, such as pseudogap phase, giant magneto-resistance, high-temperature superconductivity etc. [3]. Secondly, it provides us means to extract technologically relevant responses from these materials, especially in this age of micro- and nano-electronics. Understanding strongly correlated electron physics opened up doors to new technologies. One such example is Giant Magneto-Resistance (GMR), which revolutionized the field of data storage [4]. The phenomena strongly correlated electrons harbor are also extremely important from a technological perspective. For example, nickelates host a highly tunable metal-insulator transition which can be used to create next generation energy efficient computer architecture known as neuromorphic computing [5]. These computing platforms provide an effective solution to scalability of energy-hungry artificial intelligent systems. High temperature superconductors host the key to the future in terms of realizing practical quantum computation, faster transportation, and many more [6]. Colossal magnetoresistance in manganates [7], metal-insulator transition in vanadates show incredible promise for next-generation quantum devices [8]. Thus, understanding these phenomena in strongly correlated electron systems will not only benefit the scientific motivation but also facilitate technological aspirations.

1.2 Common Themes of Strongly Correlated Electron Physics

Explaining phenomena in strongly correlated electron systems is probably one of the fundamental pursuits of present condensed matter physics research efforts. As a result, phenomena in transition metal oxides such as cuprates, nickelates, vanadates, manganates, ruthenates, etc. are typically studied. It is usual to study the response of the system as a thermodynamic parameter other than temperature, say if γ is varied. Typically, for a given value of the thermodynamic parameter, the system is cooled down to lower temperatures, where the collective order of the constituents changes to indicate a phase transition. As we discussed, the real space of the ground state is textured with hosting homogeneous regions of ordered ground state, possibly separated from each other by high temperature phase of the material. In what follows, we highlight two specific themes that are common in most strongly correlated electron physics.

Phase Texture

In such strongly correlated electron systems, one of the main themes is the complexity and organization of structures [3]. A common characteristic of the phase diagrams of various complex oxides is that different low temperature ground states emerge from a general high temperature state upon the judicious varying of a tuning parameter, represented schematically in Fig. 1.1. This leads to the competition and/or coexistence between phases where neither of the two low temperature phases is strongly favored and are nearly degenerate. Furthermore, often at these border regions of parameter space, where both phases are degenerate, there is a local tendency toward either of the phases but globally, neither of the phases dominate. Thus, a metastable, laterally textured pattern emerges across the sample with lengths on the order of nanoscale to mesoscale. Also, at these border regions between phases, one usually finds exotic quantum ground states (e.g., charge density waves [9]) and unusually large responses to small external perturbation (e.g., colossal magneto resistance [3]). A typical example of a metastable state is found during first-order phase transition, where the system exhibits hysteresis. For such metastable states, it is important to characterize the heterogeneity locally. For example, in the nickelates, various recent reports, using x-ray photoelectron emission microscopy (X-PEEM), have observed a preferential nucleation and growth of insulating domains in a background of metallic

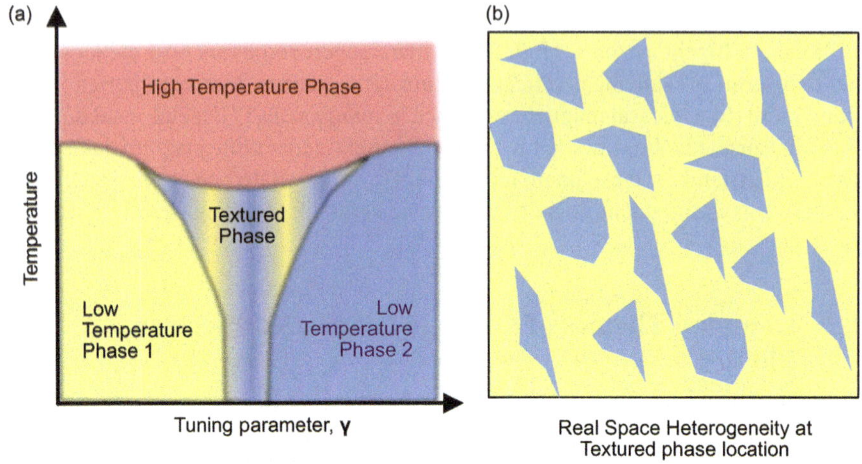

Fig. 1.1 Phase competition and phase coexistence in strongly correlated electron systems: **a** As a function of the tuning parameter a metastable textured phase emerges for in-between values where neither phase is preferred for a low temperature ground state phase. **b** Schematical representation of a textured phase

regions along the surface morphology indicating the active role of the structure on the phase separation [10]. New imaging techniques deployed on an $NdNiO_3$ film to visualize textured metal-insulator state found the coexistence of a first- and second-order phase transition phenomena at the domain boundaries which was attributed to coupling of charge and spin order parameters [11]. Another astounding report using a nano-focused x-ray beam tuned to the Ni L3 edge to detect the Bragg peak associated with the antiferromagnetic order noticed fractal geometric arrangement of antiferromagnetic domains hinting toward a proximity to a critical point [12]. In the manganates, disorder-driven heterogeneous texture of two competing phases of ferromagnetic metal and antiferromagnetic insulator is observed when colossal magneto resistance emerges and the two are argued to be related to each other both experimentally and theoretically [13]. The vanadates also feature remarkably inhomogeneous Mott insulating phases. At the onset of the first-order transition, the rutile metallic phase coexists with an insulating monoclinic phase, forming non-trivial domain textures. These domains have recently been observed to form filaments upon the application of electric field pulses, that leads to resistive switching [14]. Near these exotic quantum phases and meta-stability dominated phase space, whether the emergence of lateral heterogeneity is a cause or an effect of exotic quantum states, remains unanswered (Fig. 1.2).

We will also indulge in a material that shows a technologically relevant phenomenon—electrically tunable switching of resistance in the region of phase space of a highly tunable antiferromagnet, where two low temperature ground states compete with each other [15], creating a state of the material which hosts meta-stable regions of textured ground states. Another useful way to think about the phenomena

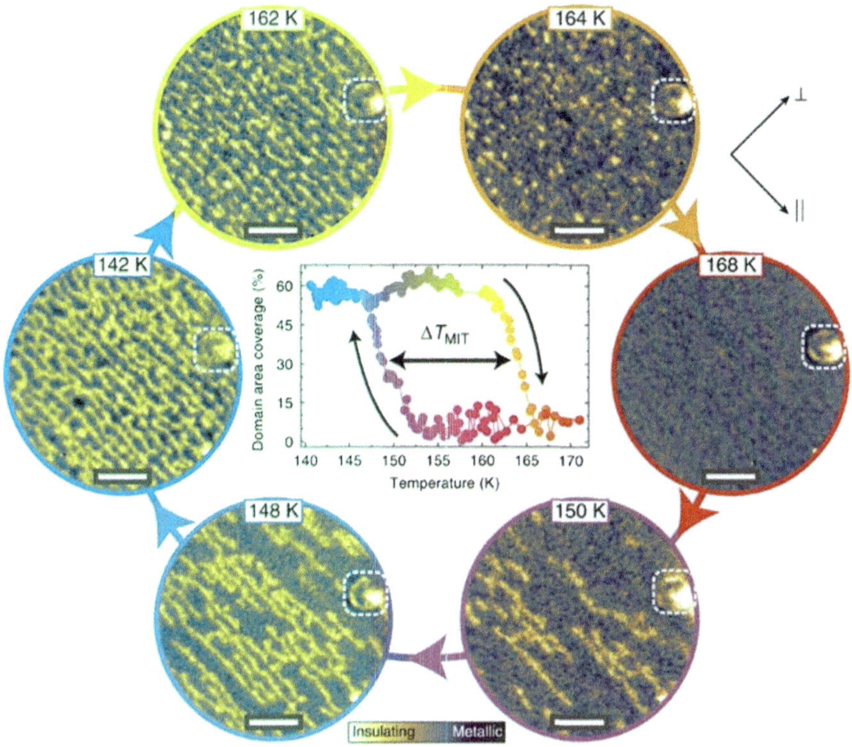

Fig. 1.2 Real space texture evolution during a first order phase transition: Here we have reproduced the result of Mattoni et al. where the authors have shown the heterogeneity of electronic state in $NdNiO_3$ films. Measurements were done using X-ray photoelectron emission microscopy (X-PEEM) [10]. Throughout the thesis for simulation purposes we have used their measurements as masks to create hypothetical real space structure

of pattern formation is to consider the phase diagram at fixed temperature while we vary the tuning parameter, γ. Since at certain γ the phase competition and phase texture evolution appear for the same temperature, it may host signatures related to the quantum effects in the material. Furthermore, the domain evolution and domain wall fluctuation are inherently long wavelength phenomena hosting extremely low energy physics. Thus, we conjecture that studying the textures—their evolution and fluctuation under γ—quantum properties of the material can be studied which were previously overlooked. Here we intend to develop the tools necessary to study the patterns of relevant quantum phases in strongly correlated quantum materials.

Antiferromagnetism

Furthermore, a typical strongly correlated electron system is realized in Mott insulating paradigm, which is defined as a strong nearest neighbor Coulomb interaction regime, which prohibits even a half-filled electron band to be itinerant. In such Mott insulators, the electrons have to pay a high price in energy if they want to hop to the

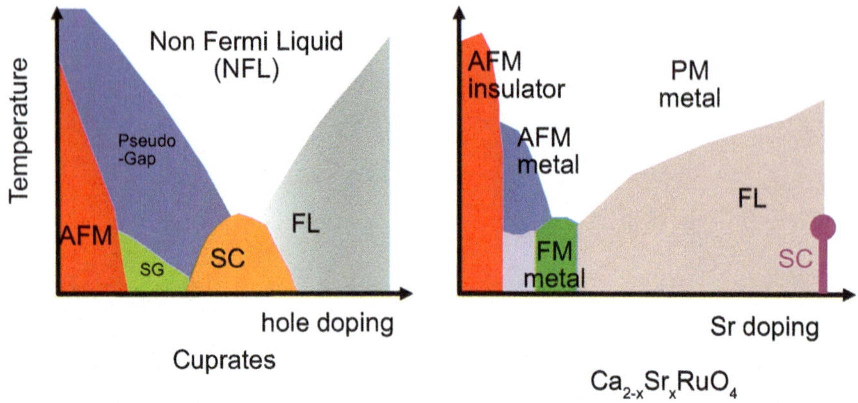

Fig. 1.3 **Representative phase diagrams**: Out of many correlated oxides here we show two phase diagrams highlighting the proximity of the Antiferromagnetic(AFM) phase to complex electronic phases such as pseudo-gap phase, splin glass (SG), antiferromagnetic metal etc. adapted from [3]

nearest neighbor site to doubly occupy a site. However, the desire of electrons to hop and lower their kinetic energy gives rise to an interesting phenomenon of antiferromagnetism [16]. In some insulating materials, electrons from the half-filled shells of cations, find intermediate coupled state with the non-magnetic anions. Through the anion, the electrons hop from cation to cation to lower its energy. This mechanism is known as the super-exchange. The high-energy interaction between cation and anion can be integrated out in favor of a low-energy description of the spin in terms of Heisenberg spin model where the nearest neighbor spin Hamiltonian is described as

$$H_{AFM} = J \sum_{\langle ij \rangle} \mathbf{S_i} \cdot \mathbf{S_j} \tag{1.1}$$

Such a super-exchange interaction usually results in a J>0 effective description of the spin Hamiltonian if the half-filled cations overlap with the same anion orbital. Pauli's exclusion principle protects the super-exchange interaction and enforce J>0. In such a scenario, the energy is further lowered by the nearest neighbor cation spins anti-aligning with each other. Thus, often Mott insulating strongly interacting electron systems are observed to be Antiferromagnetic. A slight de-tuning of the electronic environment from the typical Mott insulating state results in these novel quantum phases in the vicinity of decreasing spin correlation (for example, in Fe-based pnictides, Cuprates etc.) where hole doping takes the role of γ previously discussed. This is shown in the Fig. 1.3 with different phase diagrams of oxides showing interesting exotic phases emerge near the antiferromagnetic phase.

Thus, we see that incredibly rich physics exists near the vicinity of Mott insulating antiferromagnetic phase where one phase is not energetically favored over the other but a heterogeneous meta-stable textured phase of the material, prevail. Locally, exceptional events, such as defects, can favor one phase over the other for such strongly coupled electronic systems. Thus, the relevant electronic phase to develop the tool to study the formation and evolution of pattern formation, is clearly antiferromagnetism. Furthermore, due to a strong coupling among different degrees of freedom (such as spin and charge), by extension, different phases of the material (such as magnetic order and charge order) compete and/or co-exist with other phases. Thus, the phase organization in real space at the meso-scale gets significantly influenced by the physics of the shorter length scale in strongly correlated electron system. Even though meso-scale phase heterogeneity and evolution are such important features in these systems [17], little advancement has been made in understanding them, primarily due to the lack of proper experimental tools to tackle the problem at different length scales simultaneously.

1.3 Antiferromagnetic Textures Relevant for Technology

After discussing the relevance of studying antiferromagnetic textures for strongly correlated materials for advancement of the understanding of these complex materials, here we discuss the importance of studying the antiferromagnetic textures for technological advancements. Antiferromagnetic materials are extremely important for next-generation emerging technology of spintronics, which specialize in low-energy and high-speed information processing. Instead of the charge in electronics, spintronics utilizes the spin degrees of freedom for storing and transporting information. Antiferromagnets are particularly advantageous for such applications owing to the following key characteristics:

- they are robust against external magnetic field perturbation
- they do not produce stray fields
- they display ultrafast dynamics, orders of magnitude faster than its ferromagnetic counterpart
- they generate large magneto-transport effects for efficient read out of the information

Antiferromagnetic materials further exhibit vanishingly small dipolar interaction, making the nearest neighbors free to gain a phase. That phase term evolves slowly over a long distance, resulting in interesting textures such as domain walls, vortices, and skyrmions, interpreted as the topological defects of the order parameter field. Since the topological defects in antiferomagnets are further protected from fluctuations, they offer robust memory solutions. The wider variety of spin textures for a given ordering vector allows richer realization of different domain wall configurations. Furthermore, light-like motion of the textures limited by the magnon group velocity enables high-speed transport of robust memory extremely important for next generation computation technology. We briefly discuss a few of the antiferromagnetic textures to provide a broader overview of the richness this field offers. In an antiferromagnetic crystal, a domain is the spatial expanse over which the amplitude

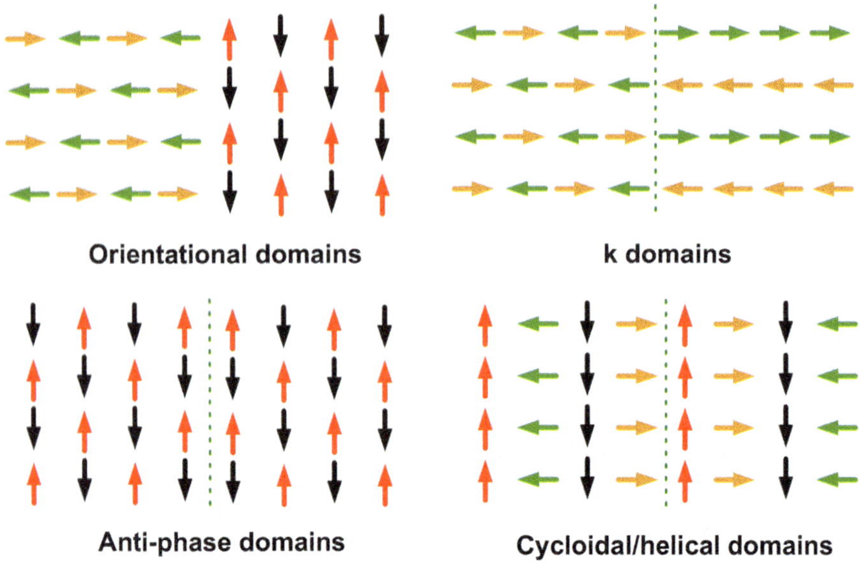

Fig. 1.4 Different antiferromagnetic domains and domain walls: Many different ways of classifying the order parameter to distinguish two domain state for antiferromagnets exist resulting into a zoo of domain structures. We show here the domains we discuss in the text after [18]

and the phase of the order parameters correspond to a given domain state. For a given crystal lattice symmetry, the anti-ferromagnetic superlattice can be realized as having different momentum vector orientation relative to the crystal lattice reciprocal space vectors. These are called k-domains. Furthermore, the relative orientation between magnetic moment directions between two domains can be different even though they follow antiferromagnetic coupling, resulting into orientational-domains. Translational domains have the same moment orientation and the same wave vector of spin modulation but differ in phase. Thus, we see that due to zero magnetization, the choice of phase can be arbitrary, resulting into complex spin textures. For instance, the phase of the spin moments can turn in a cycloidal or helical fashion. They can be colinear and non-colinear. So far we have been discussing domains that differed in one of the dimensions and allowed a domain wall in between. However, the structures separating the two domains can become more complex. For such complex spin orientations, the idea of a domain wall can be generalized in higher dimensions using the definition of topological defects of a field defined as

$$\oint_C \nabla \mathbf{M} \cdot d\mathbf{r} \tag{1.2}$$

This is known as winding number of the field and is non-zero for a loop around a topological defect. If the antiferromagnetic moment is confined to two dimensions, it results in the structures called vortex and anti-vortex. For a three-dimensional

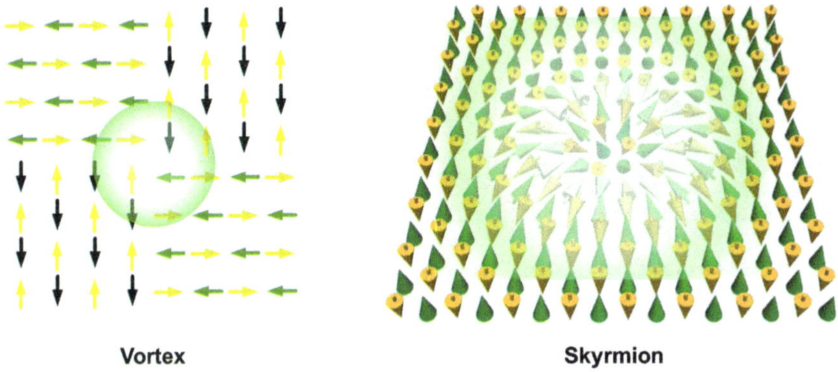

Fig. 1.5 **Vortices and Skyrmion**: (left) Schematic representation of a Z_4 vortex defect, (right) skyrmion phase [19]. The shaded region represent a loop that encompasses the singularity in the antiferrmagnetic field. Thus winding number evaluated over these paths will result into a non-zero value

spin moment, we realize skyrmion. Both vortices and skyrmion, themselves, are two-dimensional structures.

Figure 1.4 shows a Z_4 type vortex on a two-dimensional lattice. In thin films of α-Fe_2O_3, $Fe_{1/3}TaS_2$, and $Ca_2SrTi_2O_7$ observation of Z_6 AFM vortex/anti-vortex have been reported [18]. The skyrmions in antiferromagnets can be manipulated using spin-polarized currents [19]. This further opens up possibilities of studying driven-dissipative systems in the context of antiferromagnetic textures. Ideas similar to magnetic race-track memory but in the context of antiferromagnets, can leverage a higher frequency of operation along with a low error in encoding information due to stray fields (Fig. 1.5).

1.4 Experimental Methods for Studying AFM Textures

Spatial inhomogeneity can be studied fundamentally in two ways:

- Scanning mode
- Single shot mode

Scanning Mode

A typical scanning mode experiment involves rastering an atomically fine tip sensitive to electronic states over a large region of the sample to study the heterogeneity. Figure 1.6 shows one such example, where the magnetic phase heterogeneity of the material could be studied. Scanning tunneling microscopy sensitive to the gapped states had been extremely important in the study of cuprates and twisted bilayer graphene, finding out longer length scale phase organizations emerging from smaller

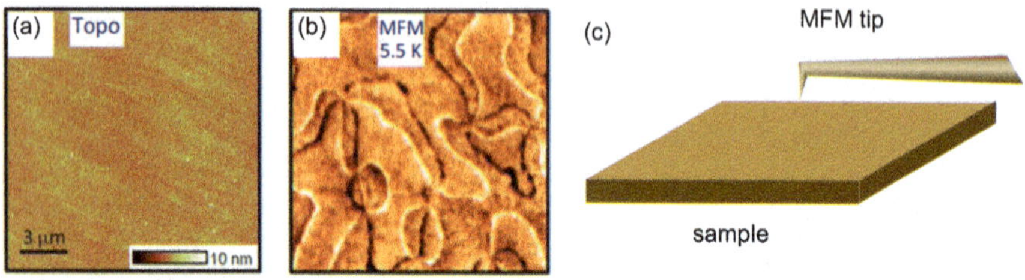

Fig. 1.6 Antiferromagnetic texture from Magnetic Force Microscopy (MFM) measurement: **a** Topography of the sample, **b** Vortex phase adapted from [20]

scale interactions. Scanning techniques for studying antiferromagnetism are prevalent and utilized for both direct and indirect probes. These techniques include resonant and non-resonant scanning magnetic diffraction, spin-polarized scanning tunneling microscopy, magneto-optical Kerr effect microscopy, magnetic force microscopy, spatially resolved second harmonic generation, and magnetic linear dichroic scanning transmission x-ray microscopy (STXM), among others. However, the method suffers from two major difficulties:

1. Since it is a scanning mode experiment at the atomic resolution it is difficult to gain statistical information about a phase organization phenomena. Thus the measurement is confined to a limited region of observation.
2. Since the atomic scale sensitivity requires high stability the measurement process is inherently slow limiting the use of it to study the evolution of the textures under perturbations. Thus the measurement can not be extended to study kinetics or dynamics of the phases.

The tool to study the electronic phase heterogeneity in complex quantum materials should be sensitive to the $\mathring{A}$-nanometer scale correlations, while being able to map the heterogeneous phase of a specific correlation at tens of micrometer length scale. Furthermore, to study the evolution of the phase texture, we need them to be faster. Thus, there is a requirement of a single-shot experiment to characterize the heterogeneity of electronic phases.

Single Shot Mode

Single-shot experiments usually utilize electromagnetic radiation over a large field of view and the sensitivity of light-matter interaction at the lowest length scale to capture the heterogeneity of phase in a system. A single-shot experiment, to discern heterogeneity, relies on a contrast mechanism. The fundamental idea about contrast is that if the sensitive region responds to one mode of incoming radiation (infrared, optical, or X-rays) and the rest of the material does not, then one can image a certain state of electronic configuration using a single shot of exposure over a large field of view. Magneto-Optic-Kerr-Effect microscopes are based on creating the contrast mechanism in optical reflection based on the polarization sensitivity of the magnetic moment. Other single-shot experiments involve X-ray photoelectron emission microscopy, which develops the contrast mechanism based on the occupied electronic levels of the system. Thus, for a given energy the photo-electron, emission will create a spatial filter for the states responding to it by ejecting photo-electrons. Those electrons are captured by a high electric field objective lens and focused to form an image. The advantage of this method relies on the direct observation of the contrast. However, the experimental method requires putting the sample under a strong electric field and high X-ray flux. Both of which can potentially damage the sample. The high flux is necessary to generate a statistically significant signal from a small field of view as big as 15 μm, with a resolution of 30 nm. X-ray magnetic linear dichroism can be used in conjunction with X-PEEM to make the experiment sensitive to image heterogeneous antiferromagnetic phase.

Another mode of single shot experiment relies on interference patterns generated by the scattered intensity to sense the heterogeneity of phase. We discuss this methodology in great detail throughout the thesis. Interference patterns are a way to generate contrast by taking the Fourier transform of sensitive regions using a coherent incoming beam. The structures of the heterogeneity leave their mark on the amplitude of the Fourier transform which we detect. Throughout the dissertation, we will discuss how to interpret such a scattering outcome. Coherent scattering is inherently a large field of view operation. We can make it sensitive to the electronic correlation by making it resonant with the electronic degrees of freedom. This requires an incredible upgrading in terms of high brilliance requirements for X-ray light sources around the world. Recently, more and more light sources are moving towards a fully coherent mode of X-ray generation tunable to a specific energy and polarization. Thus, in a timely study, we dive into developing experiments and understanding resonant coherent scattering data collected from strongly correlated electron systems.

1.5 Overview: Antiferromagnetic Textures in the Context of This Thesis

Through the investigation of antiferromagnetic materials, we analyze the textured real space caused from a strong coupling of charge and spin. We demonstrate that this strong coupling results in a textured real space in antiferromagnetic materials. Parallely, to visualize the phenomena, we develop and apply soft X-ray coherent scattering to explore these real-space textures [18]. The dissertation is organized as follows:

In Chap. 2, we introduce the method of resonant coherent soft X-ray scattering. We explain its principles and its application in imaging antiferromagnetic textures. This technique leverages the resonant enhancement of X-rays at specific absorption edges, enabling high contrast and high spatial resolution in magnetic imaging. We further illustrate the methodology, stemming from the technique, to observe nucleation and growth. Our observations and conclusions provide insights that underscore the utility of coherent X-ray scattering in studying magnetic phenomena at the nano-scale, especially in the dilute limit.

Chapter 3 extends the use of resonant coherent soft X-ray scattering. In conjunction with other spectro-microscopy techniques, we create and analyze artificial lateral heterostructures. Our novel approach of using helium nano-focused ion beam implantation to locally modify the properties of nickelate thin films, effectively engineers these artificial antiferromagnetic textures; thus modifying the local physics in the material. As we observe and control the formation of these antiferromagnetic domains, we discuss the possibility of using this methodology in tailoring a material towards targeted magnetic properties.

In Chap. 4, we explore the melting phenomena in antiferromagnets using coherent resonant soft X-ray scattering using Fe-intercalated transition metal dichalcogenides. The material exhibits a highly tunable magnetic phase diagram, providing an opportunity to study the melting process under thermal fluctuations and phase competition. Our simultaneous sensitivity to a specific antiferromagnetic order through resonance Bragg condition and the real space distribution of the ordered regions through the coherent aspect of the scattering is shown to be extremely helpful in understanding phase competition in strongly correlated magnetic systems.

Finally, in Chap. 5, we introduce a new approach to resonant coherent soft X-ray scattering by splitting a coherent beam spot into multiple beam spots. We show that our methodology offers improved data acquisition efficiency and enhanced flexibility in experimental design, as compared to traditional single-beam experiments. We discuss the technical implementation of this dual-beam approach and its potential applications in studying complex magnetic textures. The thesis concludes by proposing further advancements in coherent scattering techniques, demonstrating the potential for these methods to provide deeper understanding of the properties of strongly correlated electrons at the mesoscale [17].

References

1. Anderson PW (1972) More is different. Science 177(4047):393–396
2. Cao Y, Fatemi V, Fang S, Watanabe K, Taniguchi T, Kaxiras E, Jarillo-Herrero P (2018) Unconventional superconductivity in magic-angle graphene superlattices. Nature 556(7699):43–50
3. Dagotto E (2005) Complexity in strongly correlated electronic systems. Science 309(5732):257–262 July
4. The discovery of giant magnetoresistance (2007)
5. Zhang H-T, Park TJ, Zaluzhnyy IA, Wang Q, Wadekar SN, Manna S, Andrawis R, Sprau PO, Sun Y, Zhang Z, Huang C, Zhou H, Zhang Z, Narayanan B, Srinivasan G, Hua N, Nazaretski E, Huang X, Yan H, Ge M, Chu YS, Cherukara MJ, Holt MV, Krishnamurthy M, Shpyrko OG, Sankaranarayanan SKRS, Frano A, Roy K, Ramanathan S (2020) Perovskite neural trees. Nat Commun 11(1):2245 2020
6. Bednorz JG, Müller KA (1986) Possible high t c superconductivity in the ba- la- cu- o system. Zeitschrift für Physik B Condensed Matter 64(2):189–193
7. Ramirez AP (1997) Review article: colossal magnetoresistance. J Phys: Cond Matt
8. Hu L, Zhu XB, Sun YP (2023) Strongly correlated antiferromagnetic vanadates. APL Mat 11(9)

9. Frano A, Blanco-Canosa S, Keimer B, Birgeneau RJ (2020) Charge ordering in superconducting copper oxides. J Phys: Cond Matt 32(37):374005
10. Mattoni G, Zubko P, Maccherozzi F, van der Torren AJH, Boltje DB, Hadjimichael M, Manca N, Catalano S, Gibert M, Liu Y et al (2016) Striped nanoscale phase separation at the metal–insulator transition of heteroepitaxial nickelates. Nat Commun 7(1):13141
11. Post KW, McLeod AS, Hepting M, Bluschke M, Wang Y, Cristiani G, Logvenov G, Charnukha A, Ni GX, Radhakrishnan P, Minola M, Pasupathy A, Boris AV, Benckiser E, Dahmen KA, Carlson EW, Keimer B, Basov DN (2018) Coexisting first- and second-order electronic phase transitions in a correlated oxide. Nat Phys 14(10):1056–1061. Publisher: Nature Publishing Group
12. Li J, Pelliciari J, Mazzoli C, Catalano S, Simmons F, Sadowski JT, Levitan A, Gibert M, Carlson E, Triscone J-M et al (2019) Scale-invariant magnetic textures in the strongly correlated oxide ndnio3. Nat Commun 10(1):4568
13. Miao T, Deng L, Yang W, Ni J, Zheng C, Etheridge J, Wang S, Liu H, Lin H, Yang Yu et al (2020) Direct experimental evidence of physical origin of electronic phase separation in manganites. Proc Natl Acad Sci 117(13):7090–7094
14. del Valle J, Kalcheim Y, Trastoy J, Charnukha A, Basov DN, Schuller IK (2017) Electrically induced multiple metal-insulator transitions in oxide nanodevices. Phys Rev Appl 8(5):054041
15. Nair NL, Maniv E, John C, Doyle S, Orenstein J, Analytis JG (2020) Electrical switching in a magnetically intercalated transition metal dichalcogenide. Nat Mater 19(2):153–157
16. Goodenough JB (1955) Theory of the role of covalence in the perovskite-type manganites [la, m (ii)] mn o 3. Phys Rev 100(2):564
17. Laughlin RB, Pines D, Schmalian J, Stojkovic BP, Wolynes P (2000) The middle way. Proc Natl Acad Sci 97(1):32–37
18. Cheong S-W, Fiebig M, Wu W, Chapon L, Kiryukhin V (2020) Seeing is believing: visualization of antiferromagnetic domains. npj Quantum Mat 5(1):1–10. Publisher: Nature Publishing Group
19. Gomonay O, Baltz V, Brataas A, Tserkovnyak Y (2018) Antiferromagnetic spin textures and dynamics. Nat Phys 14(3):213–216
20. Geng Y, Lee N, Choi YJ, Cheong S-W, Wu W (2012) Collective magnetism at multiferroic vortex domain walls. Nano Lett 12(12):6055–6059

Chapter 2
Coherent Scattering of Resonant X-Rays from Antiferromagnetic Textures

Abstract A scattering, produced by phase-coherent electromagnetic radiation, is what we define as coherent scattering. A phase-coherence among different parts of the wavefront, both in the spatial and temporal sense, is achieved to make the incoming X-rays coherent. A fully coherent beam, like a laser, has a well-defined phase relationship between two sections of the beam separated in space-time. When such phase-structured light scatters from an assembly of scatterers, the information about spatial distribution of the scatterers gets embedded into the phase of the scattered beam. Now, in the far-field limit, different parts of the beam interfere with each other to give rise to an interference pattern on the detector. So in a coherent scattering experiment, the outcome is an interference pattern which is a fingerprint of the spatial distribution of the scatterers. In this chapter, first we introduce coherent X-ray scattering and then we present our work detailing the result of coherent scattering from an antiferromagnetic pattern.

2.1 Introduction and Importance of Coherent Scattering

Every coherent scattering is essentially an interference experiment where spatial coherence of the incoming beam is exploited like Young's double-slit experiment (Fig. 2.1). Before going any further into the discussion, we would like to contrast this with Michelson's interference experiment, where the temporal coherence of the incoming beam is utilized. In a Michelson interferometer, a coherent beam is split in 'time', where one part travels a reference distance and the other part goes into investigating the physics in question, and in the end, both of them are made to interfere. In such an interference experiment, the field is measured over time to extract the interference pattern. On the other hand, here a coherent portion of the beam is split in 'space', and the two sections encode their phase environments into their phase information. Finally, they are made to interfere with each other on a detector, where we observe an interference pattern in space. Unlike the Michelson experiment where field variation over time is measured, here we record spatial intensity distribution in reciprocal space. Similar to the Young's double-slit experiment, where observed fringe separation in the interference pattern is used to evaluate the structure of the

R. Basak, *Nucleation, Fabrication, and Competition in Antiferromagnetic Textures Studied with Coherent Resonant Soft X-ray Scattering*, Springer Theses,
https://doi.org/10.1007/978-3-032-19199-1_2

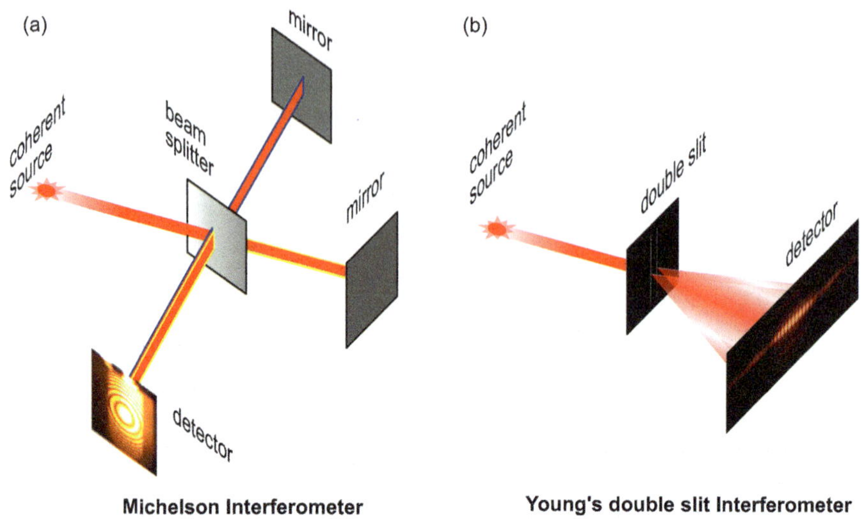

Fig. 2.1 **Schematic of two basic interference experiment**: **a** Michelson interferometer, where the incoming coherent beam is split in time (here represented with two different colors of the edges of the beam) and **b** Young's double-slit interferometer, where the incoming coherent beam is split in space. The representative detector images are taken from Wikimedia Commons. The beam split in time undergoes different path length differences before coming back together to interfere with each other. The detector images are enlarged for clarity

slit arrangement, i.e. their separation, we can extract structural information about the distribution of scatters. However, as we will see in the later parts of the thesis, often the arrangement of the scatterers is complex and thus, leads to a complex interference pattern. The complexity arises both from the number of scatterers and their relative random arrangement. The interference pattern we observe in coherent scattering experiments is known as the 'speckle pattern' from its origin in astronomy.

2.1.1 Young's Double-Slit Interference

In this section, we revisit Young's double-slit interference to introduce the ideas related to coherent scattering. First, we will describe the intensity observed as a statistical quantity and introduce ideas such as 'contrast' and 'mutual coherence function'. We will use this result to introduce the ideas such as transverse and longitudinal coherence. A schematic of the experiment is shown in the Fig. 2.2.

From the two pinholes separated by d, coherent light passes to reach the detector at a distance D on the propagation axis. Following the superposition principle, the waves are added on their amplitude basis and then squared to find the resultant intensity distribution on the intensity-sensitive detector. Since the detector time response is

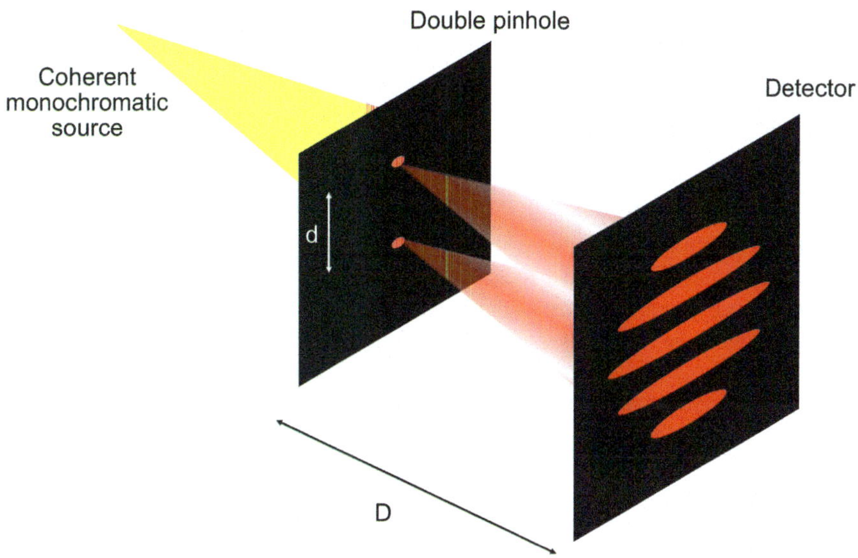

Fig. 2.2 Interference from two pinholes: Interference through two pinholes observed as a sinusoidal intensity variation on the detector. The intensity distribution in detector is measured in $I(\mathbf{q})$

usually low for photon counting operations, the quantity is time-averaged to find the intensity distribution. The spatial distribution of the time-averaged intensity is a sinusoidal function—the fringe pattern. We can express the intensity at a position $\mathbf{q}$ as—

$$I(\mathbf{q}) = \langle \psi^*(\mathbf{q}, t)\psi(\mathbf{q}, t)\rangle, \tag{2.1}$$

where the $\psi(\mathbf{q}, t)$ is the superpositioned resultant of the two individual amplitudes and $\langle .. \rangle$ represent averaging over time. Since we are interested in expressing this intensity in terms of the two pinholes, we redefine the amplitude as a weighted superposition of the individual amplitudes at those pinholes, sub-scripted by 1 and 2.

$$\psi(\mathbf{q}, t) = A_1\psi\left(\mathbf{p_1}, t - \frac{\mathbf{r_1}}{c}\right) + A_2\psi\left(\mathbf{p_2}, t - \frac{\mathbf{r_2}}{c}\right)$$

The locations of the pinholes 1 and 2 are denoted by $\mathbf{p_1}$ and $\mathbf{p_2}$. We further note that, $\mathbf{r_i} = \mathbf{q} - \mathbf{p_i}$, where $i \in 1, 2$. Here we retarded the wavefronts in time and considered a general scenario where the two signals are added with a ratio of $A_1 : A_2$ both of which are complex constants. Putting this back into Eq. 2.1—

$$I(\mathbf{q}) = I_1(\mathbf{q}) + I_2(\mathbf{q}) + A_1A_2^*\langle\psi\left(\mathbf{p_1}, t - \frac{\mathbf{r_1}}{c}\right)\psi^*\left(\mathbf{p_2}, t - \frac{\mathbf{r_2}}{c}\right)\rangle + c.c. \tag{2.2}$$

Now, $I_1(\mathbf{q})$ and $I_2(\mathbf{q})$ are the incoherent contribution from respective pinholes and equates to $I_i(\mathbf{q}) = |A_i|^2 \langle |\psi(\mathbf{p}_1, t - \frac{\mathbf{r}_1}{c})|^2 \rangle$, for $i \in 1, 2$. The cross-terms become relevant only in coherent scattering. They are called *mutual coherence function* and defined as—

$$\Gamma_{12}(\tau) = \langle \psi(\mathbf{p}_1, t + \tau) \psi^*(\mathbf{p}_2, t) \rangle \tag{2.3}$$

This is the first order in amplitude cross-correlation between the two sources. Since we propagate the wave function by a phase to evaluate the intensity at the detector, it is evident that the cross-correlation also propagates preserved and shows up in the intensity distribution at the detector. Redefining time as $t \to t + \frac{\mathbf{r}_2}{c}$ we have,

$$I(\mathbf{q}) = I_1(\mathbf{q}) + I_2(\mathbf{q}) + A_1 A_2^* \Gamma_{12}\left(\frac{\mathbf{r}_2 - \mathbf{r}_1}{c}\right) + A_1^* A_2 \Gamma_{21}\left(\frac{\mathbf{r}_1 - \mathbf{r}_2}{c}\right) \tag{2.4}$$

From (2.2) we see that the mutual coherence function has the property such that $\Gamma_{12}(-\tau) = \Gamma_{21}^*(\tau)$. Incorporating this we can simplify the the expression of $I(\mathbf{q})$ further into:

$$I(\mathbf{q}) = I_1(\mathbf{q}) + I_2(\mathbf{q}) + 2A_1 A_2 Re\left[\Gamma_{12}\left(\frac{\mathbf{r}_2 - \mathbf{r}_1}{c}\right)\right], \tag{2.5}$$

where we have also considered A_i to be purely imaginary constants. This is the central equation describing the coherent scattering. Since Γ_{12} is a function of $\mathbf{r}_2 - \mathbf{r}_1$ whose magnitude is d, we have successfully shown to capture the structural information of the sample plane in the intensity distribution of the interference pattern on the detector plane. For a purely monochromatic illumination illuminating ideal pinholes, the mutual coherence function is just a simple phase, and thus the real part of it is just a sinusoidal function. This is observed as a fringe pattern on the intensity distribution at the detector.

From this exercise, we can directly see that the inclusion of complexity comes from just increasing the number of pinholes, which in turn increases the number of fields to account for during superposition at $\mathbf{q}$. The resultant intensity will thus depend on all possible cross terms that can come from considering the superposition of multiple sources resulting in a very complex intensity distribution. Even then, the intensity distribution will only depend on the structural distribution relations of the pinholes in the sample plane. This is the primary reason for the great interest in coherent X-ray scattering. Later we will look at ways to understand this intensity distribution and how that reveals information about the spatial distribution of scatterers. For the derivation, we considered a fully coherent beam illuminating the pinholes. In the following sub-section, we will explore how the fringe visibility depends on the coherent property of the beam.

2.1.2 *Measure of Coherence*

Turning the argument upside down, fringe visibility from a double-slit interference can be used as a measure to quantify the coherence property of the illuminating beam. The coherence property of the illuminating beam is defined in terms of coherence lengths: longitudinal coherence length (ξ_L), and transverse coherence length (ξ_T), defined with respect to the beam propagation direction, $\mathbf{k}_{in}$.

Transverse Coherence

To discuss transverse coherence, we relax the condition of a point source illuminating the two pinholes, for an extended source. Different elementary areas of the extended source independently radiate wavefronts. We are interested in finding the upper bound of such an extended source, which can still produce interference fringes. Here, each will produce the interference fringes independent of The other. Fringe Visibility ($\mathbb{V}$) is defined by Michelson for a modulated intensity varying between I_{max} and I_{min} as—

$$\mathbb{V} = \frac{I_{max} - I_{min}}{I_{max} + I_{min}} \tag{2.6}$$

Clearly, the fringe visibility will wash out if the fringe pattern produced by the central emitter and the boundary emitter overlap out of phase, as schematically shown in Fig. 2.3. It is easy to evaluate for a forward scattering in a small angle approximation where the angles subtended by the source at the sample plane add up to the angles of diffraction where the central maxima will show up. In other words, a shift away from the central axis translates the fringe pattern by an angle subtended by the source on the pinhole plane. The higher order of the maxima will be separated by $\frac{\lambda D}{d}$, where λ is the wavelength of illumination, d is the slit separation, and D is the distance between the sample plate and the detector. Thus, for such an extended emitter, the condition for total loss of contrast is found.

$$\frac{w_{max} L}{2R} = \frac{\lambda L}{2d} \tag{2.7}$$

where R is the distance between the source and the pinholes. Thus, for a source with transverse dimensions below w_{max}, we will see interference fringes. The separation on the pinhole for a given dimension w is called the coherence length in the transverse direction (ξ_T). Thus, for a given source of w dimension, the coherence length is

$$\xi_T = \frac{\lambda R}{w}.$$

Longitudinal Coherence

The discussion about longitudinal coherence is where we relax the condition of monochromaticity (Fig. 2.4). $\mathbb{V}$ will reduce to zero if two wavelengths appear out-of-phase with each other. Then similar to the above situation, where the maxima

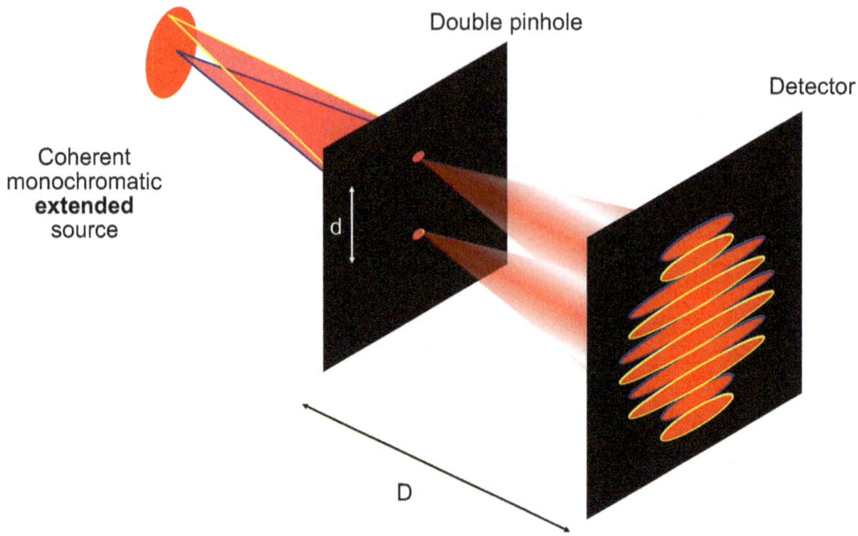

Fig. 2.3 Effect of extended source: The extension of source decreases transverse coherence. This is schematically explained in this figure where the source point away from the propagation axis can be understood as a source experiencing tilt. Then the interference pattern will be tilted in the other direction. For a tilt that shifts the central maxima of the boundary source point of the extended source by one fringe width we will lose the fringe contrast and the detector will appear uniformly illuminated. This argument is a simple way to evaluate the relation between transverse coherence and source size. However, in reality, every source points superposition together before interfering to get the overall fringe pattern. This exercise was a first order approximation for the complex process

of one wavelength appears, a minima of the other will overlap. Let us consider the spread in the wavelength as $\Delta\lambda$. Similarly, we compare the fundamental frequency (λ) and the extreme outlier ($\lambda + \Delta\lambda$ to find the coherence length. The longitudinal coherence length is defined as the length (ξ_L) such that, when the primary component oscillates N times within that length, the other one must oscillate N-$\frac{1}{2}$. Thus, solving for N from $N\lambda = (N - \frac{1}{2})(\lambda + \Delta\lambda)$ and substituting into the definition, we find,

$$\xi_L = N\lambda = \frac{\lambda^2}{2\Delta\lambda}.$$

2.1.3 Coherence in Synchrotron Radiation Sources

There are two main types of X-ray photon sources: (1), Gaussian or chaotic light sources, such as X-ray tubes, Synchrotron radiation storage rings, and (2) single-mode emission in self-amplified spontaneous emission (SASE) in X-ray free electron lasers. Here, we focus on the experiments in Synchrotron radiation sources. The ideas

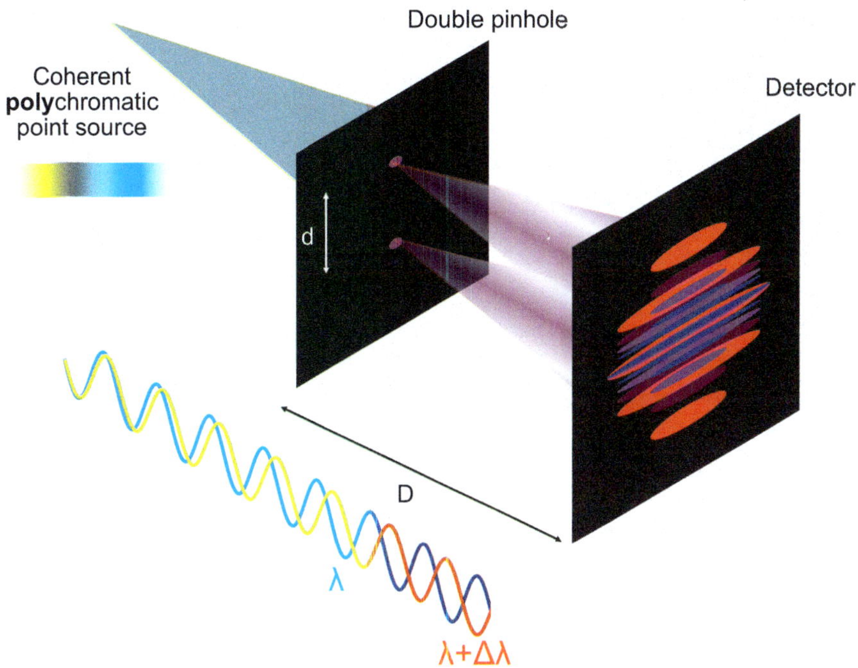

Fig. 2.4 Effect of polychromatic source: The polychromatic source decreases longitudinal coherence. The loss of fringe contrast is achieved but in a different path than the case of an extended source. We have shown for the case of a polychromatic source of 4 different wavelengths. Two extreme wavelengths are compared to assist the derivation of longitudinal coherence length (ξ_L). It can be seen that for 8 wavelengths of the blue wave, the red wave completes only 7.5

we discussed before apply directly to Synchrotron radiation sources, where multiple modes of coherent emission are present. Thus, no coherent source at a Synchrotron is fully coherent for this reason. However, we will see that a coherent portion can be extracted following the ideas developed so far.

Here we always deal with partially coherent Synchrotron radiation. We saw that limiting the spatial extension perpendicular to the propagation axis helps us improve the transverse coherent percentage of the incoming beam. This is achieved using pinholes or slits along the beam path. On the other hand, temporal filtering is done by using a monochromator to extract a small $\Delta\lambda$. A double bounce monochromator, as shown in the Fig. 2.5, is a typical choice to do the job that achieves a $E/\Delta E \sim 10^4$. However, the price we pay is a huge loss of intensity. But, if the source has enough photon flux to begin with, such as third-generation Synchrotron sources the coherent flux achieved after filtering has enough photon to get speckle patterns after scattering. Undulators, compared to bending magnets and wigglers, offer considerably more Brilliance, $\mathbb{B}$, defined as the number of photons per unit time, generated per unit source area, per unit solid angle, and 0.1% fractional bandwidth.

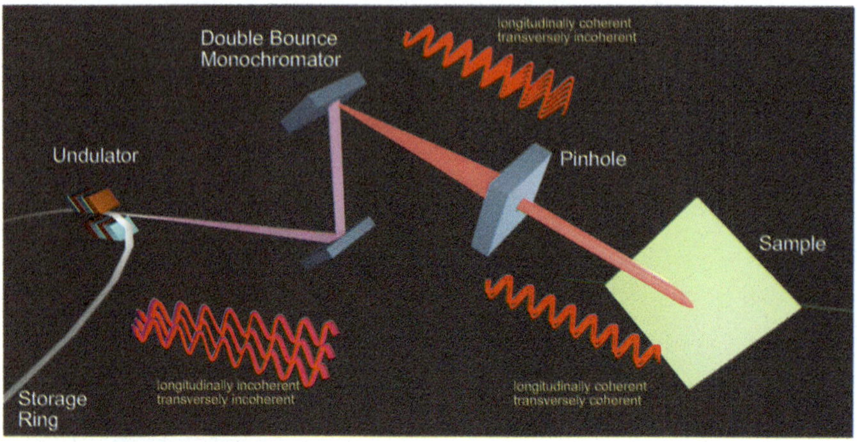

Fig. 2.5 Coherent beam formation in Synchrotron sources: Double bounce monochromators make the beam longitudinally coherent then pinhole is employed to filter out the divergent section of the beam. We schematically represent the wave fronts as the beam propagates through the instruments and gets processed to become fully coherent. While the phase relationship between different parts of the incoherent beam evolves longitudinally or transversely, for the coherent beam it remains fixed. Coherence does not require the phase to be the same for every section of the beam but to be well-defined. This is schematically represented in this figure

Measure of the Coherent Flux

Synchrotron radiation sources have a Gaussian intensity distribution. If we consider the source has a size of $\sigma_h \times \sigma_v$ in the horizontal and vertical directions, respectively. The corresponding coherence lengths are given by

$$\xi_T^a = \frac{\lambda R}{2\pi\sigma_a},$$

for $a \in h, v$. For a typical source size of $\sigma_v \approx 10 - 50\,\mu\text{m}$ and $\sigma_h \approx 100 - 500\,\mu\text{m}$, at a wavelength of 1 Å, and a typical distance of 40m away from the source, the transverse coherence area is 3–10 μm × 25–100 μm.

Now, let us estimate the photon flux through such a single coherence area spanned by $\xi_h \times \xi_v$. If the source has a brilliance of $\mathbb{B}$ the photon flux, F, transmitted by the coherence area is

$$F = \mathbb{B} \times 4\pi^2\sigma_h\sigma_v \times \frac{\xi_T^h\xi_T^v}{R^2} \times \frac{\Delta\lambda}{\lambda}$$

which reduces to

$$F = \mathbb{B}\lambda\Delta\lambda \tag{2.8}$$

Thus, we find that the key quantities of the source longitudinal coherence length, transverse coherence area, and flux through the coherence area depend primarily on the brilliance $\mathbb{B}$, the wavelength of choice λ, and the bandwidth of the monchromator

$\Delta\lambda$. A typical Si(111) monochromator has a bandwidth on the order of 10^{-4} which gets us a longitudinal coherence length of 400 nm and a flux on the order of 10^{11} photons.s^{-1}, sufficient for resonant scattering investigations in condensed matter systems, provided we have a Bragg peak to scatter off from.

2.2 Coherent X-Ray Scattering

2.2.1 Principles of Coherent Scattering

Coherent illumination of an aperture function results in an intensity distribution on the detector plane that is related to the aperture function through the Fourier transform. In this section, we will go through the basic result of the Rayleigh–Sommerfeld diffraction solution in scalar diffraction theory. The result we obtain will be used in interpreting the detector speckle pattern.

Here the field distribution in the source plane is given by $\psi_s(\xi, \eta)$. Following the Huygens–Fresnel principle of wave propagation, each point on the aperture plane will act as a secondary source emitting a spherical wave with wave vector **k**, where $|k| = \frac{2\pi}{\lambda}$ and λ is the monochromatic coherent beam illuminating the aperture plane. Let us consider that both the detector plane and the aperture plane are normal to **k** and their separation is R. Then the instantaneous intensity at the (q_x, q_y) in the detector plane can be found from the superposition of all spherical waves—

$$\Psi(q_x, q_y) = \frac{\iota}{\lambda} \int_A \psi_s(\xi, \eta) \frac{e^{\iota \mathbf{k} \cdot \mathbf{r}}}{r} d\xi d\eta \tag{2.9}$$

From the figure we can write $r = \sqrt{r_0^2 + (q_x - \xi)^2 + (q_y - \eta)^2}$ where r_0 is considerably larger than other terms. Thus, we can utilize binomial expansion to express them as

$$r = r_0 + \frac{\xi^2 + \eta^2}{2r_0} - \frac{q_x \xi + q_y \eta}{r_0} \tag{2.10}$$

As R increases, so does r_0. Since ξ and η are finite, the contribution from the curvature term decreases with increasing R. Let us consider that $\xi, \eta < a$, then for a separation beyond $r_0 \sim R > \frac{a^2}{\lambda}$ this term contributes unity. Such a separation separates the Fresnel diffraction regime from the Fraunhofer diffraction regime. The curvature of the spherical wavefronts remains important in the superposition in the Fresnel diffraction regime where $R < \frac{a^2}{\lambda}$. In Chap. 5, we will utilize the expression including the curvature and the nuances it brings into the observed intensity distribution. However, here we focus on the Fraunhofer regime. Furthermore, in this far-field regime, **k** and **r** are approximately parallel, yielding $\mathbf{k} \cdot \mathbf{r} = kr$. Thus, with these considerations, the equation appears as follows:

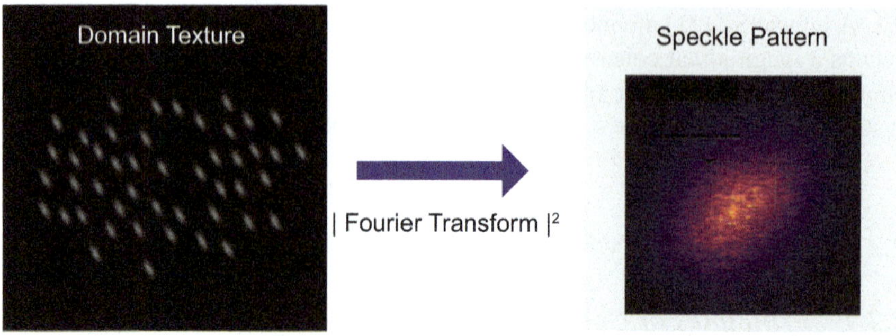

Fig. 2.6 Speckle pattern is the fingerprint of domain texture: The real space texture of an electronic order is related to the speckle pattern by an amplitude Fourier transform. Since the detector is insensitive to phase information we only retrieve amplitude information as depicted in this schematic

$$\Psi(q_x, q_y) = \frac{\iota}{\lambda} e^{\iota k r_0} \int_A \psi_s(\xi, \eta) e^{-\iota \frac{k}{r_0} q_x \xi} e^{-\iota \frac{k}{r_0} q_y \eta} d\xi d\eta \tag{2.11}$$

This is just the Fourier transform of $\psi_s(\xi, \eta)$. Thus, in the far-field limit, the intensity distribution ($I(q_x, q_y)$) can be found:

$$I(q_x, q_y) = \Psi^* \Psi = |\mathcal{F}(\psi_s)|^2 \tag{2.12}$$

This is the result we set out to show. The immediate consequence observed is that the phase information of the complex field ψ_s is lost. This is known as the phase problem in coherent scattering. However, a considerable amount of research has shown that under certain circumstances iterative algorithms can retrieve the phase. However, this falls outside the scope of this thesis. Interested readers may consult these papers for reference (Fig. 2.6).

2.2.2 Coherent X-Ray Scattering in Practice

So far we have shown that coherent scattering measures the mutual coherence function at the detector plane, and the resultant intensity distribution is given by the Fourier transform of the scatterers in the aperture plane. However, to produce a coherent outcome on the detector we need to scatter from within the coherence volume of the X-rays generated in a high-brilliance Synchrotron source. In this section, we will consider further experimental considerations for a scattering experiment.

First, we will note the trivial consequence of transverse coherence length. This limits the largest size of the object (let's call this dimension, a) studied under a single shot of exposure. Next, we consider the consequence of longitudinal coherence for an experiment, where momentum transfer $\mathbf{q}$ happens with $\mathbf{q} = \mathbf{k}_{out} - \mathbf{k}_{in}$, where $\mathbf{k}_{out}$ and $\mathbf{k}_{in}$ are the wave vectors of the scattered and incoming beam. This is often the case we will explore. In such a situation the path length difference between the waves from the center of the sample and the edge of the sample should be below the longitudinal scattering length, or

$$\frac{\lambda^2}{\Delta\lambda} > asin(2\theta) \tag{2.13}$$

Relating $q = 2ksin(\theta)$ we can rewrite the equation as

$$\frac{\lambda^2}{\Delta\lambda} > a\frac{q}{k}\sqrt{1 - q^2/4k^2}, \tag{2.14}$$

and follow by neglecting higher order terms in $\frac{q}{k}$ as

$$\frac{\lambda^2}{\Delta\lambda} > a\frac{q}{k}. \tag{2.15}$$

Thus, if $q = \frac{2\pi}{d}$ where d denotes the d-spacing of the lattice planes in question then we find

$$\frac{\Delta\lambda}{\lambda} > \frac{d}{a} \tag{2.16}$$

Thus, for a given coherent diffraction experiment investigating a coherent intensity pattern on the Bragg peak with a defined d-spacing the maximum object size investigated under a single shot is limited by the above equation. Typically the d-spacing associated with a Bragg peak is on the order of few Å, and the relative energy resolution ($\Delta E/E$) of a monochromator is typically on the order of 10^{-4}, yielding the transverse coherence length, a to be on the order of several μm.

2.2.3 Coherent Resonant X-Ray Scattering

In what we discuss further in this thesis, it will involve doing a resonant scattering experiment in the coherent mode. Incorporating resonance scattering with a coherent source opens up interesting possibilities and challenges. Furthermore, the experiments we perform are mostly done in the soft X-ray regime. Firstly, in resonance conditions, scattered intensity is enhanced tremendously, which compensates for the relatively low coherent flux for larger wavelengths in soft X-rays following the equation of F.

Secondly, the experiment benefits from a longer coherence length. However, the penetration depth of the beam remains sufficiently small at the resonance condition. Since the top few hundred nm of the sample absorb the incoming photon with a high cross-section, the coherent volume remains very small yielding high contrast compared to non-resonant scattering experiments. Furthermore, if we are exploring an order with a correlation length larger than the penetration depth of the coherent beam the scatter distribution can be considered effectively two dimensional. That we can neglect the thickness of the sample for a small thickness of scattering volume, is called the projection approximation.

Furthermore, short scattering length due to resonance significantly decreases the chances of multiple scattering. On top of this when we are investigating the electronic order the light-matter interaction is already very weak. We enhance the cross-section significantly by using resonance conditions. However, it is still a very low-intensity peak compared to any structural Bragg peak. Thus, resonant coherent scattering is inherently photon-hungry. In a nutshell, the weak scattering, projection approximation, and hence, the lack of multiple scattering makes the coherent scattering in resonance conditions a high contrast coherent investigation. Due to such added advantages, experiments of coherent imaging are possible in coherent scattering in the soft X-ray limit.

To carry out imaging experiments, the soft X-ray beam needs to be focused down to a tight spot. However, the focusing optics, such as zone plates, are difficult to work with in the soft X-ray regime. At soft X-ray wavelength, the focal length is extremely low sometimes within 1–2 mm from the sample surface, where an order sorting aperture needs to be fit. This challenge is, however, avoided if the experiment can be performed using a pinhole. Most of the experiments in this thesis utilize pinholes of diameter 10 μm.

The main incentive to conduct coherent resonant scattering lies in the physics investigated. Through resonance conditions, the incoming beam is sensitive to electronic domains and textures. In strongly correlated electron physics, multiple degrees of freedom couple to each other. Order formation in one degree of freedom such as, say spin, affects the energy landscape of the other degrees of freedom, for example, the charge, and as a result, co-existence and competition of orders emerge. The study of coherent resonant scattering is fundamentally the study of the emergence, growth, coexistence, competition, and melting of electronic orders. We will explore them time and again within the context of the thesis.

2.3 Correlation Analysis

In the previous section we have shown that a mutual coherence function is the cross correlation between two amplitudes and shows up as intensity modulation on the detector. In this section, we investigate the correlations that survive if, instead, we correlate two intensity patterns. This is modeled after Hanbury–Brown–Twiss interferometry, where intensity measured on two separate detectors is cross-correlated. Here we will not use two different detectors, however, we will correlate two instances of the same detector (Fig. 2.7).

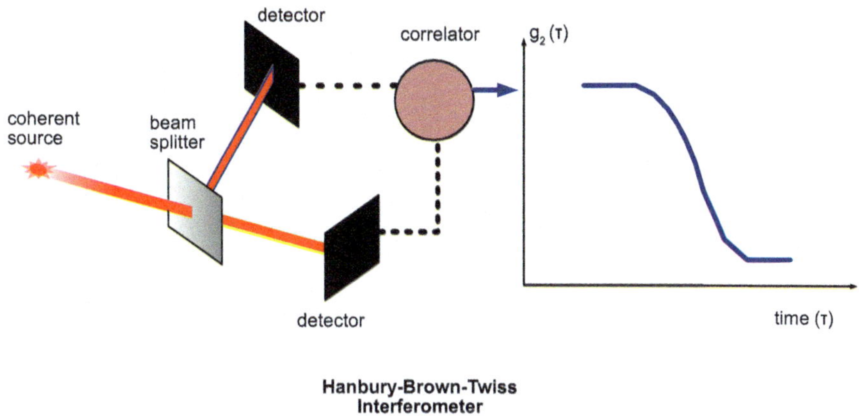

Fig. 2.7 **Intensity-intensity correlation**: The correlation analysis is based on the Hanbury–Brown and Twiss interferrometer where the intensities observed in two detectors are correlated to evaluate the cross correlation. This is the governing principle for speckle correlation analysis between two instances of intensity distribution on the detector

Hanbury–Brown and Twiss's experiment measured temporal coherence of the second order. A quasi-monochromatic light from scatterers was incident upon a 1:1 beam-splitter, where half intensity was directed towards one detector, and the other half to the second detector. The two detectors are photon counters, and they measure the intensity on the detector surface. The two measured intensities are then time-delayed by τ before being correlated. This is called two-time auto-correlation.

$$g^{(2)}(\tau) = \frac{\langle I(t)I(t+\tau)\rangle_t}{\langle I(t)\rangle_t^2} \tag{2.17}$$

In contrast to first-order correlations, where scattered field amplitudes from the sample are correlated, here we find the second-order correlations correlating the intensities. As we found previously, the mutual coherence function is a two-point correlation finding correlation between points at $\mathbf{r}_2$ and $\mathbf{r}_1$. Following the same argument, the intensity correlation in its full generality is a four-point correlation. Writing it at its most generality, we find,

$$g^{(2)}(\mathbf{q}_1, \mathbf{q}_2, \tau) = \frac{\langle \Psi(\mathbf{q}_1, t)^* \Psi(\mathbf{q}_1, t) \Psi(\mathbf{q}_2, t+\tau)^* \Psi(\mathbf{q}_2, t+\tau)\rangle_t}{\langle I(\mathbf{q}_1, t)\rangle_t^2} \tag{2.18}$$

where τ is the time separation between two measurements and $\mathbf{q}_1$ and $\mathbf{q}_2$ are two points on the detector. In this general definition of two-point two-time cross-correlation, we have set up a relationship among the scattered fields observed on the detector. In condensed matter systems, we study $g^{(2)}$ between two measured intensities taken at two instances of a parameter. Often that parameter is the time,

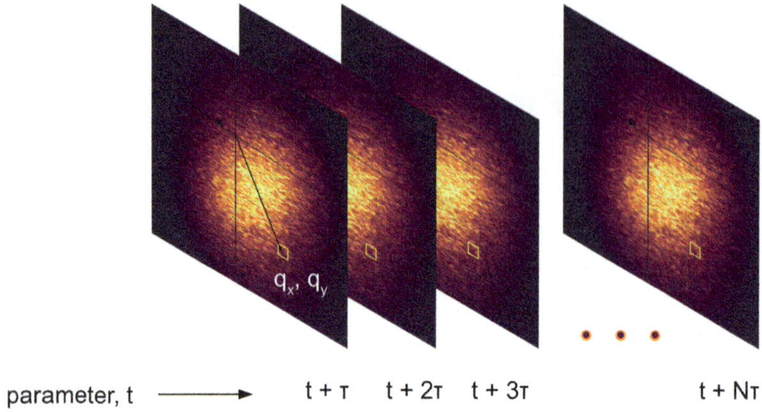

Fig. 2.8 Intensity series as a function of a parameter, t: Measurement of intensity speckle pattern as a function of a parameter (can be time, temperature, external field, or triggering events such as current pulses) as it is varied by discrete steps of τ

temperature, or magnitude of an external field which changes the domain distribution and hence the speckle pattern on the detector. Experimentally, the correlation function is evaluated by the convolution of two speckle patterns observed at two different parameter values. Furthermore, often in practice, we evaluate the speckle pattern evolution at a given $|\mathbf{q}|$, or a region of interest (ROI) in the detector space with $(\Delta q_x, \Delta q_y)$ width at a position (q_x, q_y). Schematically such a measurement is shown for a parameter t in the figure. Many instances of the detector intensity distribution are measured and recorded. Then the intensity at (q_x, q_y) can be cross-correlated for each frame. Equation 2.18 becomes simpler for the present situation where

$$\mathbf{q}_1 = \mathbf{q}_2 = \mathbf{q} = (q_x, q_y)$$

and the $g^{(2)}$ function can then be written for the parameter t as—

$$g^{(2)}(\mathbf{q}, \tau) = \frac{\langle \Psi(\mathbf{q}, t)^* \Psi(\mathbf{q}, t) \Psi(\mathbf{q}, t+\tau)^* \Psi(\mathbf{q}, t+\tau) \rangle_t}{\langle I(\mathbf{q}, t) \rangle_t^2} \tag{2.19}$$

Such cross-correlation functions are measured during the experiments and used to find the evolution of heterogeneity as a function of the parameter. $g^{(2)}$ as defined above evaluates the average change observed by the correlation function as a function of the parameter difference τ. The cross-correlation between any two different images can be plotted for all intensity patterns recorded as well. Such a measure of correlation is not averaged over the parameter variable but is usually averaged over sample positions for a better signal-to-noise ratio and is defined formally as (Fig. 2.8)—

$$g^{(2)}(\mathbf{q}, t_1, t_2) = \frac{\langle \Psi(\mathbf{q}, t_1)^* \Psi(\mathbf{q}, t_1) \Psi(\mathbf{q}, t_2)^* \Psi(\mathbf{q}, t_2) \rangle}{\langle I(\mathbf{q}, t) \rangle^2} \tag{2.20}$$

The $\langle ... \rangle$ is taken over sample positions which is equivalent to the parameter space average for an ergodic system but different for a non-ergodic system. The measure of the two-time correlation function $g^{(2)}(\mathbf{q}, t_1, t_2)$ can measure the equilibrium and non-equilibrium fluctuation. While the correlation function $g^{(2)}(\mathbf{q}, \tau)$ measures the equilibrium fluctuation of a system. Such equilibrium fluctuation can happen between two or more equivalent energy states the system can traverse in its configuration space. Thus, the $g^{(2)}(\mathbf{q}, \tau)$ function can be related to the dynamical structure factor of a system, $S(\mathbf{q}, \omega)$, which captures the essence of the statics (distribution of constituents to form equilibrium structure) and dynamics (fluctuation of the structure around equilibrium position) of a structure in the reciprocal space. The dynamic structure factor can also be measured in an inelastic scattering experiment. However, the range of energy scales to be accessed is different for the two methods [1]. The inverse Fourier transform in frequency (ω) of the dynamic structure factor is related to the time evolution of the structure and is known as the intermediate scattering function.

$$S(\mathbf{q}, \omega) = \frac{1}{2\pi} \int_{-\infty}^{\infty} e^{-\iota\omega t} \langle \rho(\mathbf{q}, 0), \rho(-\mathbf{q}, t) \rangle,$$

where ρ denotes the spatial Fourier transform of the charge density scattering the X-rays at a given time.

X-ray photon correlation spectroscopy measures the photon fluctuation at a given $\mathbf{q}$, following the four-point correlation function. Usually, dynamic signatures of scatterers are probed in this experiment, probing energy scales of fluctuations and correlated motions. In the context of condensed matter systems, we can probe domain fluctuations at slow time scales using X-ray photon correlation spectroscopy. Van-Cittert-Zernicke theorem relates the four-point correlation function to the two-point correlation function $g^1(\mathbf{q}, \tau)$, which is defined as follows:

$$g^{(1)}(\mathbf{q}, \tau) = \frac{\mathbf{1}}{\mathbf{N}} \langle \rho(\mathbf{q}, \mathbf{0}), \rho(-\mathbf{q}, \mathbf{t}) \rangle$$

Under the ergodic approximations and short-ranged fluctuations, the four-point correlation relates to the two-point correlation as—

$$g^{(2)}(\mathbf{q}, \tau) = 1 + \beta g^{(1)}(\mathbf{q}, \tau), \tag{2.21}$$

where β is the contrast of the speckle pattern measured from Eq. 2.6. The decay of the two-point correlation can be modeled via $(\mathbf{q}, \tau) = e^{-\frac{\tau}{\tau_f}}$, where τ_f is the time-constant of the fluctuation as well as the decay of the correlation. Time correlation experiments are done in hard and soft X-ray scattering, and are called X-ray Photon Correlation Spectroscopy [1, 2]. Plethora of experiments have been able to study the change of time scales in equilibrium fluctuations of domains of the electronic orders, and have been able to connect it to the fluctuation of electronic phases and phase transitions [3–7].

Similar to the correlation function between any two time of the evolution, we can define correlation function, which can be used to find correlation between speckle pattern measured at two different temperatures T_1 and T_2. In that case, it will be called two-temperature correlation function $g^{(2)}(\mathbf{q}, T_1, T_2)$. We will use this correlation between two different temperatures to find similarity index between two speckle patterns of a melting antiferromagnetic phase. Such an experimental characterization is also used by the authors to study the texture evolution of charge density wave order as a function of temperature [8]. Similarly, suppose a system is textured with ordered domains, that couple to a thermodynamic force, such as strain, temperature, or field. We can evaluate two instances of speckle pattern as we vary the external thermodynamic force and found the correlation among the patterns. This finds the response of the texture to the external thermodynamic perturbation and many interesting physics can be investigated utilizing this method. A comprehensive reference can be found in the work of Chesnel et al., where the authors studied ferromagnetic domain evolution under external field and temperature [9].

Since the detector we measure is an area detector we can define regions of interest around $\mathbf{q}_1$ and $\mathbf{q}_2$ for no time delay, i.e. $\tau = 0$. This is utilized in finding angular cross-correlation from the scattered intensity. This is essentially symmetry in the auto-correlation function, which can be used to investigate angular symmetries for short-range correlations. If the system had long-range correlation these structures would have given a Bragg peak. However, due to short-range correlation, they are found as a ring in the q-space. Evaluating the correlation in this ring helps us find the hidden order in the apparent disorder of the material [10–13]. The intensity-intensity angular correlation term relating to the angular symmetry of the real texture is defined following Eq. 2.20 as—

$$C(q, \Delta) = \frac{\langle I(q, \phi) I(q, \phi + \Delta)\rangle_\phi}{\langle I(q, \phi)\rangle_\phi^2} - 1 \tag{2.22}$$

In practice, we subtract the incoherent background from the raw speckle pattern to extract the purely coherent signal. We find the correlation between two coherent speckle patterns through convolution. The correlation measured in this way can be related to the two-point correlation and hence structural properties of the material. We discuss extracting correlation function and similarity of speckle pattern in details, later in Chap. 4. In the following section, we will show how the speckle pattern emerges during the phase transition of an antiferromagnetic order. The order in question is first order in nature, and at the onset of the transition antiferromagnetic domains nucleate in a meta-stable energy landscape. We will show how the method of resonant coherent scattering is sensitive to the spatial arrangement of nucleating domains.

2.4 Results: Nucleation of Antiferromagnetic Domains

The correlated antiferromagnetic ground state observed in rare-earth nickelates is of interest both fundamentally, as well as, for technological application. In the past years, the non-collinear antiferromagnetism observed in $LaNiO_3$, $PrNiO_3$, and $NdNiO_3$ has been studied extensively, both in thin films [14–18], as well as, in multilayer heterostructures [19–22]. Here, we investigate the antiferromagnetic transition in a 40 nm $PrNiO_3$ thin film using coherent x-rays tuned to the $2p \rightarrow 3d$ electronic transition of the Ni ions. This electronic resonance is referred to as the Ni L_3 resonance, and it can be exploited to embed valuable information in the x-ray scattering cross section about the valence $3d$ electronic states and, for example, the antiferromagnetism which they host. Transverse and longitudinal x-ray beam coherence lengths on the order of micrometers enable sensitivity to mesoscopic real space structures with comparable length scales.

Figure 2.9a shows a schematic of the resonant coherent x-ray diffraction (RCXD) experimental setup. The orientation of the sample, and position of the detector are chosen with respect to the incidence beam such that the antiferromagnetic Bragg peak at momentum transfer $\mathbf{q}_{\mathrm{AFM}} = (\frac{1}{4}\ \frac{1}{4}\ \frac{1}{4})_{\mathrm{pc}}$ is recorded on the detector.

A 40 nm thin film of $PrNiO_3$ was grown by pulsed laser deposition from a stoichiometric target using a KrF excimer laser 2 Hz pulse rate and 1.6 Jcm^{-2} energy density. The film was deposited on a (001)-oriented $[LaAlO3]_{0.3}[Sr_2AlTaO_6]_{0.7}$ (LSAT) substrate in a 0.5 mbar oxygen atmosphere at 730 °C. The film was then annealed post-growth in a 1 bar oxygen atmosphere at 690 °C for 30 min. We note that although the LSAT substrate is known to undergo a structural transition near 150 K [23], this transition does not lead to a diffraction peak at the Q-vector corresponding to $(\frac{1}{4}\ \frac{1}{4}\ \frac{1}{4})$ in pseudocubic reciprocal lattice units of the $PrNiO_3$ film.

DC electrical resistivity measurements were performed in the 4-point van der Pauw geometry using a Quantum Design Physical Properties Measurement System. Growth and characterization of the film were performed at the Max-Planck-Institute for Solid State Research in Stuttgart. The temperature dependence of the diffraction intensity is plotted in Fig. 2.9b and compared with the DC electrical resistance of the film. As expected in this system, both signals are hysteretic, indicative of the first-order nature of this phase transition. The resonant coherent x-ray diffraction experiments were performed at the coherent x-ray scattering (CXS) beamline of the NSLS-II synchrotron light source at Brookhaven National Laboratory, which provides the coherent high-photon flux necessary to observe antiferromagnetic domains with sub-micrometer length scales. Immediately before reaching the sample, a 10 μm laser ablate pinhole aperture selects the most coherent portion of the beam and defines the probe size. The longitudinal coherence length was approximately 1.25 μm, that is about one order of magnitude longer than the soft x-ray path length through our thin film. The measurements were performed using π-polarized incident photons, with an energy bandwidth of ~0.5 eV centered at the Ni L_3 resonance (~853.4 eV). In order to access the first-order antiferromagnetic Bragg peak of the Ni spin spiral, the sample was mounted on a $\chi = 55°$ wedge so as to bring the pseudocubic (pc)

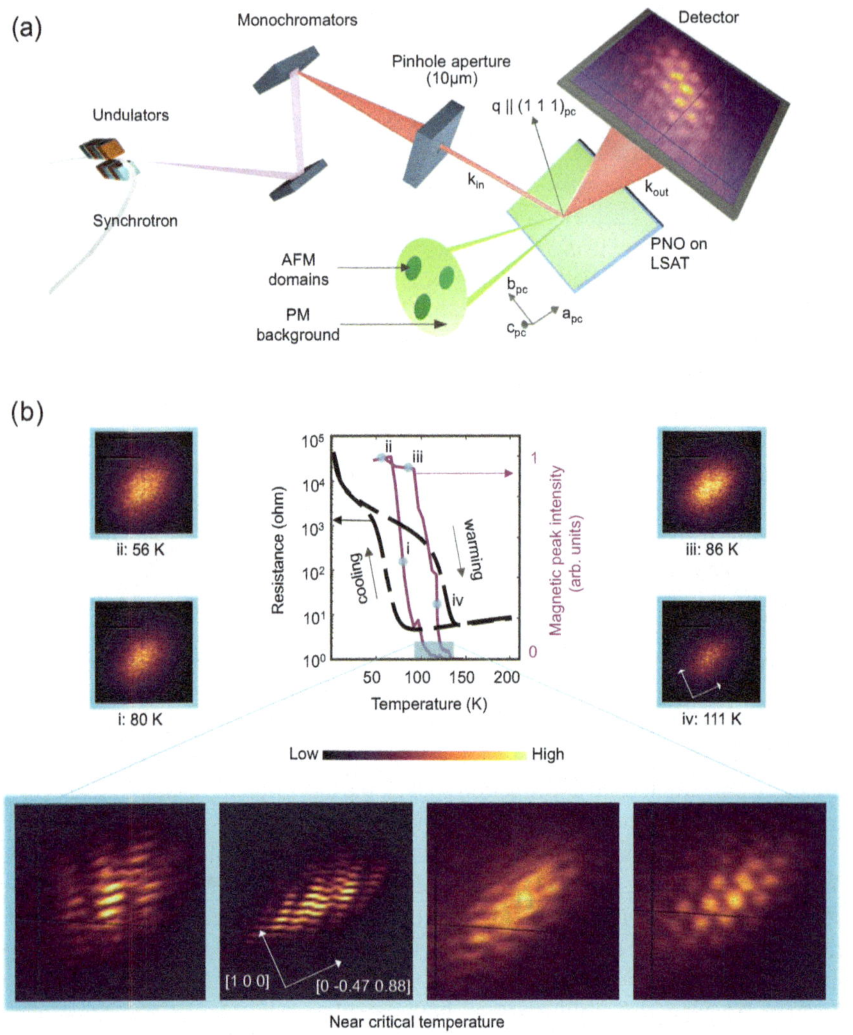

Fig. 2.9 Observation of small-q modulation on the anti-ferromagnetic Bragg peak: **a** Experimental set-up for coherent resonant X-ray diffraction **b** Resistance versus temperature plots overlayed with magnetic peak intensity evolution shows first order nature for the coupled transition. Speckle pattern at different point on the magnetic order evolution plot is shown

[111] direction into the vertical scattering plane. The $\mathbf{q}_{\mathrm{AFM}} = (\frac{1}{4}\ \frac{1}{4}\ \frac{1}{4})_{\mathrm{pc}}$ momentum transfer was then reached by going to the approximate scattering angles $\theta = 55°$ and $2\theta = 110°$. The scattered photons were detected using a fast charge-coupled device (CCD) camera, which descriminates neither the polarization nor the energy of the scattered photons. The reported intensities thus correspond to an integration over

both the π and σ outgoing polarization channels, as well as over both elastic and inelastic scattering processes.

The data shows a complex speckle pattern well below the transition temperature of $T_{Néel}$. However, the surprising feature of our data is observed for temperatures near the critical temperature of the antiferromagnet [highlighted in Fig. 2.9b]. In this temperature regime, the antiferromagnetic diffraction peak is modulated by a small-**q** superstructure. This is in contrast to the typical speckle pattern [24] which is observed at lower temperatures, where the insulating antiferromagnetic phase is well established. In Fig. 2.9b, examples of the magnetic Bragg peak at various temperatures are used to demonstrate the evolution of the diffraction pattern through one full cooling and warming cycle. The white arrows indicate the reciprocal space vectors on the detector plane.

Due to the coherence of the x-ray beam, interference between x-rays scattered at mesoscopic distances from one another (up to micrometers) is capable of producing a modulation of the Bragg peak intensity in reciprocal space. These modulations encode information about the relative sizes, positions, shapes, and the relative phases of the domains from which the x-rays are scattered. In the limit of many domains distributed across the entire beamspot, the resulting modulations are complex and the speckle size is inversely related to the size of the beam [25]. Such a typical speckle pattern is observed on the low temperature antiferromagnetic Bragg Peak in Fig. 2.9b. However, in the limit of very few scatterers occupying only a portion of the beamspot, well-defined oscillations of the Bragg peak intensity are observed in reciprocal space, and can be directly interpreted in terms of the real space distances between domains or domain boundaries. Hence, our study represents the first observation of a small-**q** modulation of an antiferromagnetic Bragg peak associated with the mesoscopic domain arrangement of the antiferromagnetic order. In our experiment, the small-**q** superstructure observed near the critical temperature is interpreted in terms of interference between the very first few domains to undergo the antiferromagnetic transition. The appearance of bubbles of antiferromagnetically ordered spins on the background of a disordered spin lattice is consistent with the bulk bimodal nature of the metal-insulator transition observed in the closely related material $NdNiO_3$ [17]. Our imaging experiment convincingly proves the first-order nature of the magnetic phase transition in $PrNiO_3$ as the domains nucleate at the onset of the phase transition.

In an RCXD experiment, although the phase of the incident light is well defined, the phases of the scattered x-rays are lost at the detector, where the collected intensities correspond to the absolute value squared of the spatial Fourier transform of the electronic states selected in the resonant process. This is known as the so-called phase problem, which implies that the real-space arrangement of the scatterers cannot be directly retrieved by applying an inverse Fourier transform to the recorded diffraction pattern without some additional information about the phase of the scattered x-rays (or preexisting information about the real space arrangement of the scatterers). The phase problem can be overcome either using holography approaches, which make use of a reference beam with known phase to determine the phase of the scattered light [26], or ptychographic [27] and coherent diffractive imaging methods [28], which are based on iterative reconstruction algorithms [29, 30]. As shown

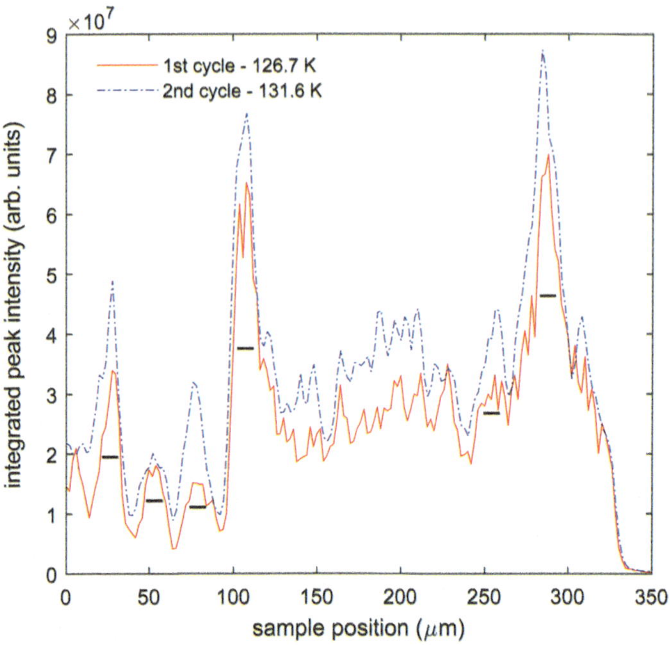

Fig. 2.10 Magnetic peak intensity as a function of beam position: The integrated antiferromagnetic Bragg peak intensity recorded on the CCD is plotted as a function of the relative beam-sample position. The two curves were taken upon warming during separate cycles through the antiferromagnetic-paramagnetic transition. The black bars indicate the width of the beam (10 μm), and highlight features whose width corresponds approximately to one beam width. The sample position scale has been given an arbitrary starting point

in Fig. 2.10, the integrated peak intensity falls off within the width of the beam, indicating well-separated distribution of domain-clusters. Thus, for the purpose of our proof-of-principle, we focus explicitly on the limit of very few scatterers, which we will refer to as the "dilute" coherent scattering regime. Working in the dilute regime, the small-**q** modulations of the antiferromagnetic Bragg peak can be interpreted by eye and modeled by trial and error based on knowledge of the system. Due to the heightened likelihood of a unique solution to the phase problem in 2 dimensions [31, 32], one can be reasonably confident that the model, which reproduces the observed diffraction pattern, is correct, thereby circumventing the need for a complex phase-retrieval algorithm.

2.4.1 Modeling the Small-q Modulation

Symmetry: To model the small-**q** modulations of the antiferromagnetic Bragg peak, we take the absolute value squared of the Fourier transform of small numbers of staggered antiferromagnetic domains, whose spatial arrangement is chosen to reflect the symmetries of the modulation patterns. We show here two cases where the symmetries were easy to identify. For a complex symmetry one can try taking inverse Fourier transform of the detector image to reveal the Patterson function—which is nothing but the auto-correlation function of the real space domain distribution. From there one can identify the principle separation of domains and their symmetries in mesoscale domain distribution [33].

Dilute cluster of domains: By translating the beam across the sample we were able to identify the observed coherent diffraction patterns as location specific, disappearing when the beam is translated by approximately one beam width. Based on this observation, we exclude repeating structures with the same symmetry as the determined domain arrangements. These are the crucial apriori knowledge we implemented to solve the phase problem. As we will show later during the modeling section, these information were crucial to reach uniqueness. In fact, a domain configuration with the minimum number of domains needed to reproduce the observed modulation pattern is consistent with the approximate space filling of the ordered phase, which is small at the onset of the transition. We first show the measurement of the integrated intensity of the peak when we scanned the beam on the sample. Diluteness of the domain is visible from the sparse high-intensity points with a spread equivalent to the beam spot size indicating a smaller spread of domain arrangement.

First order character: Guiding the modeling process in this way quickly produces a simple real space image of the relative positions and sizes of the first-ordered domains formed in the illuminated region of the film. Note that this level of analysis does not capture the detailed shape of the domains, which are given an ellipsoidal profile in order to reproduce the observed peak elongation. In general, any first-order phase transition with some degree of spatial inhomogeneity is expected to exhibit a dilute set of nucleation sites in some temperature regime near the ordering temperature. In the current system, cycling of the temperature through the phase transition reveals that the first antiferromagnetic domains formed in the nucleation regime reappear at the same locations, thereby indicating that the seeding centers are likely pinned to structural features that are immobile for temperatures up to the maximum cycling temperature 300 K. Indeed, the substrate may play an important role in determining this spatial inhomogeneity, for example, via inhomogeneous strain fields, structural domain patterns, or step edges of the substrate's terraced surface, all of which may locally suppress or seed the metal-insulator transition and accompanying antiferromagnetism. This interpretation is further supported by the observation that the same location-specific patterns are recurrent after warming to temperatures well into the paramagnetic phase. The experimentally observed RCXD pattern for 3 beam positions on the sample are shown in Fig. 2.11. The first column shows the diffraction patterns as observed upon warming during a first cycle through the transition. The remaining columns show diffraction patterns observed upon warming during the second cycle (approximately 24 h later and after warming to 300 K) and for a series of temperatures in the vicinity of the antiferromagnetic transition. Looking carefully at these diffraction patterns, it is clear that the same basic features of the small-$\mathbf{q}$

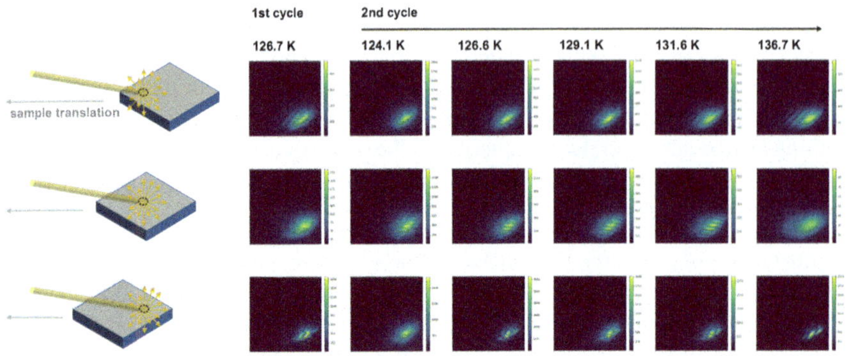

Fig. 2.11 Position and temperature dependence of the small-q superstructure: Antiferromagnetic RCXD patterns are recorded in the near-critical regime of the antiferromagnetic transition in $PrNiO_3$. Data is presented for 3 non-overlapping sample positions. Each row corresponds to a unique sample position, whereas each column corresponds to a fixed temperature value from either the first or second cycle, as labelled. The second cycle through the transition was performed approx. 24 h after the first cycle, and after warming to 300 K and cooling to 30 K in between

modulations are retrieved when returning to the same sample position and approximately the same temperature, even after fully cycling through the entire transition. At the lowest temperatures shown, the diffraction patterns begin to change due to the sudden appearance of additional ordered domains within the beamspot.

Mathematical Model

In a resonant x-ray scattering measurement of the first-order scattering from the antiferomagnetic Bragg peak, the scattered intensity is given by [34]

$$I(\mathbf{q}) = \sum_{domains} |\sum_{i} e^{\iota \mathbf{q}\cdot\mathbf{r_i}} F_i(\hbar\omega)(\varepsilon \times \varepsilon' \cdot \hat{\mathbf{m}}_\mathbf{i})|^2, \tag{2.23}$$

where $\mathbf{q}$ is the momentum transfer, r_i is the position of the magnetic moment $\hat{\mathbf{m}}_\mathbf{i}$, $\varepsilon(\varepsilon')$ is the polarization of the incoming (outgoing) photon and $F_i(\hbar\omega)$ captures the photon energy dependence of the scattering tensor element selected by ε and ε'. The scatterers are not labeled with respect to the domains they are housed. For incoherent scattering, each domain contributes to the detected intensity as a sum. On the other hand, for resonant coherent scattering from the same magnetic arrangement of scatterers is given by—

$$I(\mathbf{q}) = |\sum_{i} D(\mathbf{r_i}) e^{\iota \mathbf{q}\cdot\mathbf{r_i}} F_i(\hbar\omega)(\varepsilon \times \varepsilon' \cdot \hat{\mathbf{m}}_\mathbf{i})|^2, \tag{2.24}$$

The function $D(\mathbf{r_i})$ takes on values between 0 and 1, and is introduced to describe the antiferromagnetic domain configuration. $D(\mathbf{r_i}) = 0$ corresponds to a disordered spin and $D(\mathbf{r}_i) = 1$ to an anitferromagnetically ordered spin. The sum runs over all spin lattice sites i inside the (coherent) beamspot. Due to the coherence of the beam, scattering from all domains within the beamspot produces mutual interference, which gives rise to a small-$\mathbf{q}$ modulated power spectrum in the dilute limit.

To model the observed small-$\mathbf{q}$ modulations of the antiferomagnetic Bragg peak intensity, we have taken the two-dimensional discrete Fourier transform of a distribution of Gaussian enveloped ordered regions

$$D_j(\mathbf{r}) = e^{\frac{-(\mathbf{r}-\mathbf{r}_j)^2}{2\xi_j^2}}, \tag{2.25}$$

each representing one domain with width ξ_j and center $\mathbf{r}_j$, such that the total domain configuration D in Eq. 2.24 is given by $D(\mathbf{r}) = \sum_j D_j$. Domain distributions, which produce satisfactory fits to the data were determined by trial and error, guided by the symmetry of the small-$\mathbf{q}$ modulation and the fringe separation, and using $\mathbf{r}_j$ and ξ_j as free parameters.

Overall peak shape and corresponding domain-shape: Experimentally, the antiferromagnetic Bragg peak envelope is observed to be extended in one direction. This elongation may arise from a combination of effects including the finite size of the antiferromagnetic domains along the film growth direction, as well as a possible intrinsic anisotropy in the antiferromagnetic correlation length in the nucleation regime. Since the peak elongation is indicative of an asymmetry in the shapes of the individual domains, rather than containing information about the mesoscopic arrangement of these domains with respect to one another, determining its precise origin falls outside the scope of this study. To capture the peak elongation in our simulations, we have separated the domain width ξ_j into two widths $\xi_j^{||}$ and $\xi_j^{\perp}$, parallel and perpendicular to the axis of the elongation, where the ratio $\xi_j^{\perp}/\xi_j^{||}$ is fixed for all domains. In the coordinate system defined by the elongation direction, the corresponding expression for the elongated domain profiles is given by

$$D_j = e^{-\left(\frac{(r^{||}-r_j^{||})^2}{2\xi_j^{||}} + \frac{(r^{\perp}-r_j^{\perp})^2}{2\xi_j^{\perp}}\right)}. \tag{2.26}$$

The results of our modeling process are demonstrated for two diffraction patterns in Fig. 2.12. The Fourier transforms of the model domain configurations agree well with the measured diffraction patterns, even without the application of a numerical fitting algorithm to optimize the fits. The additional background intensity and reduced speckle contrast observed in the experimental data may arise in part from short range antiferromagnetic fluctuations, which are not captured in our model. We emphasize that, although, the solution to the real space domain arrangement remains ambiguous with respect to the trivial operations translation, phase-shift, and inversion-conjugation, the 2-dimensional nature of the probed domain arrangement allows us to exclude more complex alternate solutions [31, 32]. While inversion of our coherent scattering data using an iterative phase retrieval algorithm was beyond the scope of this study, our results imply that the numerical inversion of such reso-

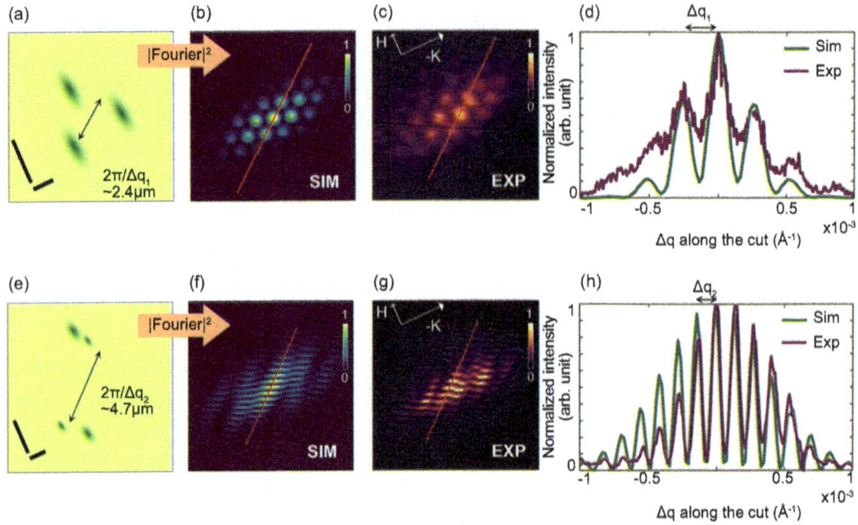

Fig. 2.12 **Modelling of experimental RCXD patterns**: **a** Model configuration in real space of three antiferromagnetic domains within the beamspot. The scale bars correspond to 2 μm along the [100] direction (long bar) and along the [010] direction (short bar). **b** Absolute value squared of the Fourier transform of the model shown in (**a**). Scale bars are the same as in (**c**). **c** Experimentally observed reciprocal space map of the antiferromagnetic Bragg peak at momentum transfer $\mathbf{q}_{\mathrm{AFM}} = (\frac{1}{4}\,\frac{1}{4}\,\frac{1}{4})_{\mathrm{pc}}$ in the pseudocubic setting of $PrNiO_3$. After projecting the detector image onto the $H-K$ plane, the lengths of the arrows labelled H and K correspond to 0.001 $Å^{-1}$ respectively along the (100) and (010) directions. **d** Comparison of cuts taken from (**b**) and (**c**) along the indicated red line. A constant has been subtracted from the experimental curves such that the intensity at the detector edge is zero. **e**–**h** same as **a**–**d** but for a different experimentally observed diffraction pattern (**g**) and the corresponding model and simulation in (**e**) and (**f**)

nant coherent antiferromagnetic diffraction data is possible in the dilute limit, and will likely be capable of both resolving finer domain morphology features, as well as extracting information from the complex speckle diffraction patterns observed at low temperatures, provided that the necessary sampling conditions are observed [35].

2.4.2 Trials of Different Domain Arrangement to Reach Solution

In this section we show the different domain arrangement we tried to reach the solution demonstrated in Fig. 2.12. We were guided by the modelling principles discussed in the previous section. In phase retrieval problems these are known as a-priori knowledge. Furthermore, these trials will provide insights about understanding speckle patterns from dilute domain distribution. Hence, what we show below forms the basis of creating a dataset to train a neural network to reconstruct the real space

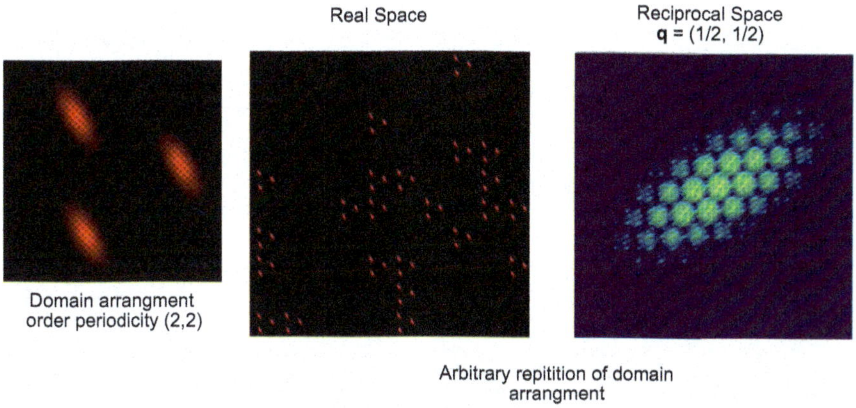

Fig. 2.13 Arbitrary arrangement of multiple domain motifs: In the real space we have repeated the motif multiple times by tiling the space with a rectangular area hosting the 3 domains. The experimentally observed pattern appears with sharp peaks on the periphery

from the dilute coherent scattering in the reciprocal space. For the simulation we are using a checkerboard pattern mimicking the antiferromagnetic order. The checkerboard pattern has an unit cell of 2×2 and thus the peak shows up at $(\frac{1}{2}, \frac{1}{2})$ position (Fig. 2.13).

Next, we show covering the real space with the motif in a few different ways. The intention is to see how this changes the speckle pattern. Interestingly, it is hard to get rid of the principle small-q modulation if there are a few domain arrangements, which gives rise to this. Thus, for the dilute-ness argument, it was essential to solve the diffraction pattern uniquely (Fig. 2.14).

We see that the small-q modulation is distorted with different random arrangements of it. If there are random domains, it does not host any small-q modulation. A curious question can be asked: how many randomly organized domains are required to completely wash off the intensity from a organized domain, giving the small-q symmetrical modulation. This is what may have happened as we lowered the temperature—eventually the nice speckle pattern was replaced with a complex speckle pattern shown in Fig. 2.15.

So far we had been playing with the multiplicity of the domain arrangement which gives rise to the experimentally observed speckle pattern. We also checked relative positions and orientation of domains, along with relative strength of the magnetic order, orientation, complex phase with respect to each other. One by one, we show below each one of the these simulations.

First, we will study the effect of geometrical transformation on the relative domain arrangement in Figs. 2.16, 2.17 and 2.18. We will show how this changes the motif, and consequent detector image for a random distribution of these motifs under the beamspot.

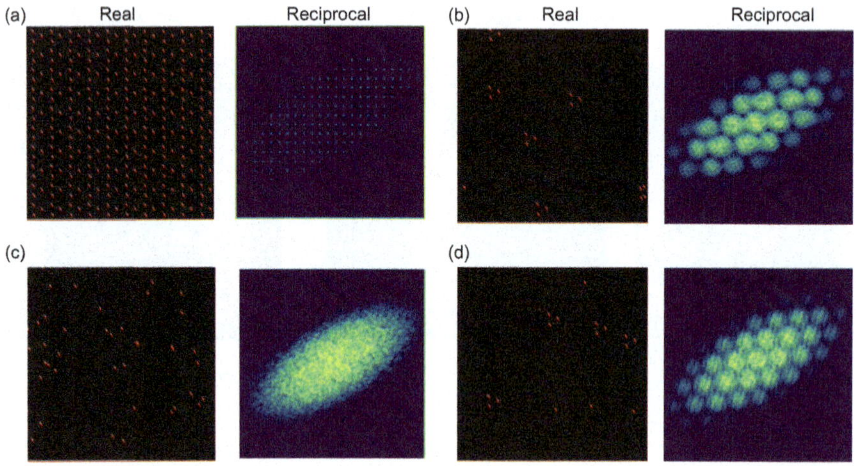

Fig. 2.14 Arbitrary arrangement of multiple domain motifs: **a** Completely filling the space with tiles hosting three domains makes the speckle pattern very different. **b** Instead of tiling the space here we place the three domain motif at arbitrary positions. In other words the separation between two domain motifs is not a multiple of a length-scale. **c** Instead of the domain motif giving rise to the experimentally observed speckle pattern we place domains randomly. **d** Mix of randomly placed domains with a few domain motifs randomly organized

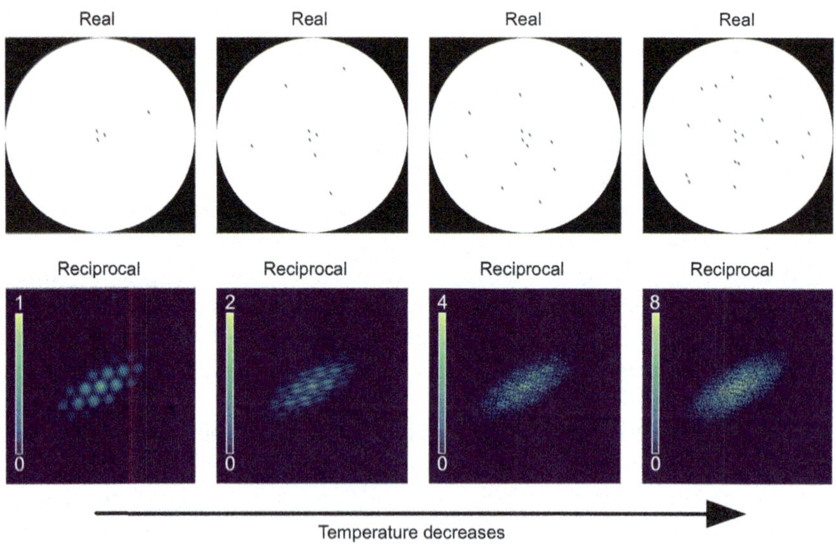

Fig. 2.15 Temperature evolution—recovering the complex speckle pattern: This is a simulation to show how many random domains are required to wash the small-q modulation off the peak. As the temperature is lowered this happens in our experiment. From left to right the temperature decreases yielding more domains to pop up under the beam-spot depicted by the white circular region in the real space. The detector space for corresponding real space is shown below

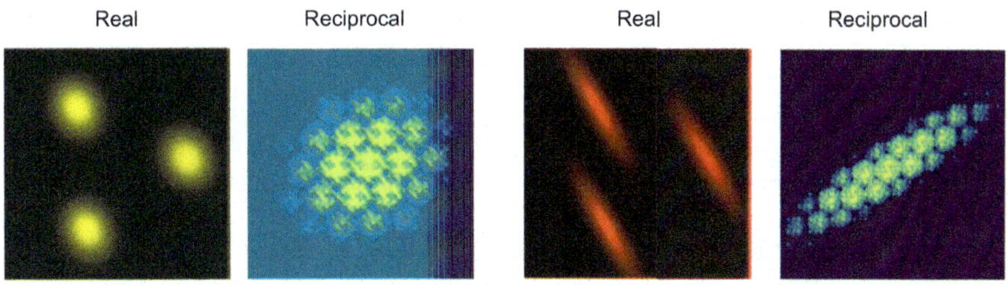

Fig. 2.16 Changes in the shape of the individual domains in the motif: The changes in the shape of the domains are directly related to the envelope of the peak

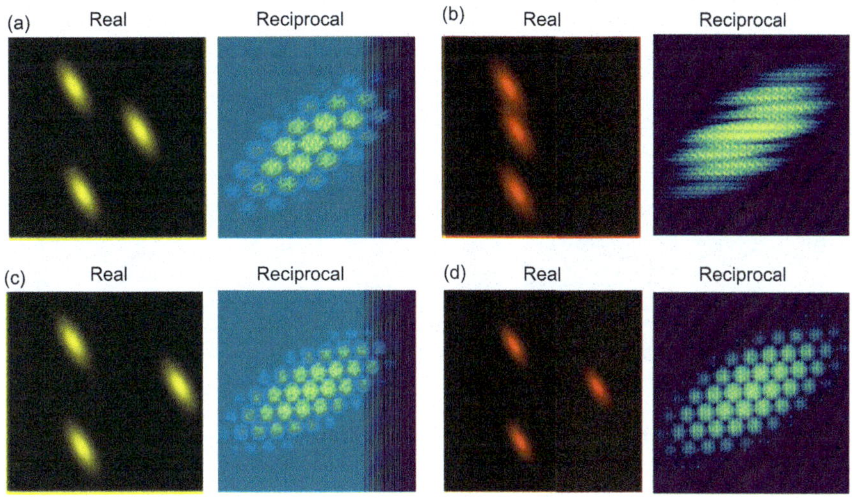

Fig. 2.17 Changes in relative position of the constituent domains of the motif: **a** The right domain is closer to the other two domains, **b** The right domain is farther to the other two domains, **c** The right domain is in-line with the other two domains **d** All three domains are separated by longer mutual separation (by 20%) as they appear smaller

Next, we will study another relative degree of freedom these domains share: amplitude and phase. Setting up the study for amplitude is trivial: we just multiply the gaussian envelope by a scaling factor. However, setting up the relative phase among domains is a bit tricky with interesting consequences. We will set up our way of solving this and the subsequent simulation results (Fig. 2.19).

Before showing the simulations of working with the relative phase, we will first go through the intuition of interference from phase. Simply put, as we have known from Young's double-slit experiment, a relative shift of π in phase will result in destructive interference. We demonstrate this in terms of domains in 1-dimension in the following figure. The key observation is that the incoherent maxima and the coherent maxima shifts relative to each other when there is a phase shift. Here, we see that if the relative phase between two real space domains differ by π, the intensity

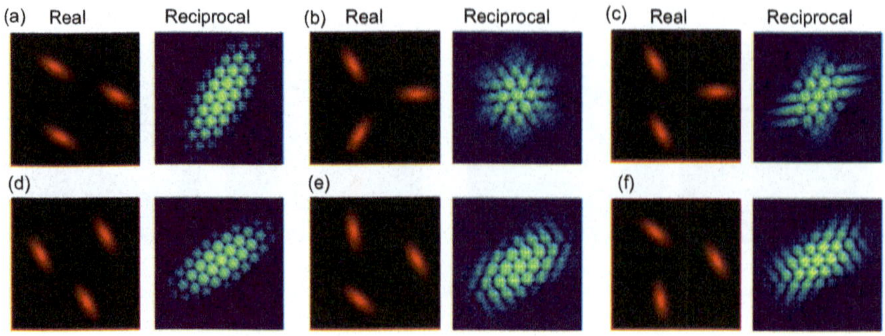

Fig. 2.18 Changes in relative orientation of the constituent domains of the motif: global rotation of the domain arrangements by 30° (**a**), and −10° (**d**); relative rotation on individual domains creates long Δq features which depend on the real space. Rotation applied on **b** left-bottom (−60°) and right domain (+60°), **c** right domain (+60°), **e** top-left (−20°) and bottom-left (+20°) domains, **f** top-left (+20°) and bottom-left (−20°) domains

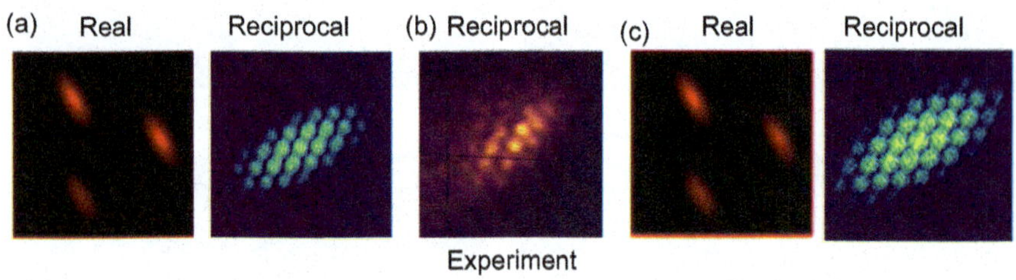

Fig. 2.19 Changes in relative amplitude of the constituent domains of the motif: **a** Only the bottom-left domain is multiplied by a scaling factor of 0.5 **b** Experimental observation similar to the intensity distribution seen in these simulation. Solving this will take adjusting the global orientation. **c** Scaling factor for the right domain 0.8, and for the bottom-left, 0.6

minima of a coherent scattering, in-line with destructive interference line-up with the incoherent peak position. We used the ideas developed in the mathematical modeling section calculate these simulations. The reciprocal space is just the amplitude of Fourier transform of the real space. A continuous sinusoidal pattern is multiplied with a Gaussian envelop in real space concurrent with the separation mentioned in the figure. The relative phase of π is implemented by multiplying -1 after the Gaussian envelope is fixed (Fig. 2.20).

With this intuition now we will go back to our 2-Dimensional simulation. Here, we will create relative phases between the three domains we have been using so far, and use the same principle to create the phase difference among them. We will introduce a phase fault between the repeating textures so that at the location of the domains we have incorporated a phase difference ($\Delta\phi$). This repeating texture, like a checkerboard, defines the unit cell and gives rise to a Bragg peak. In the following, we create a checkerboard pattern of 4 white cells and 4 black cells. A one-cell thick fault line will, thus, introduce a phase difference of $\pi/4$ (Fig. 2.21).

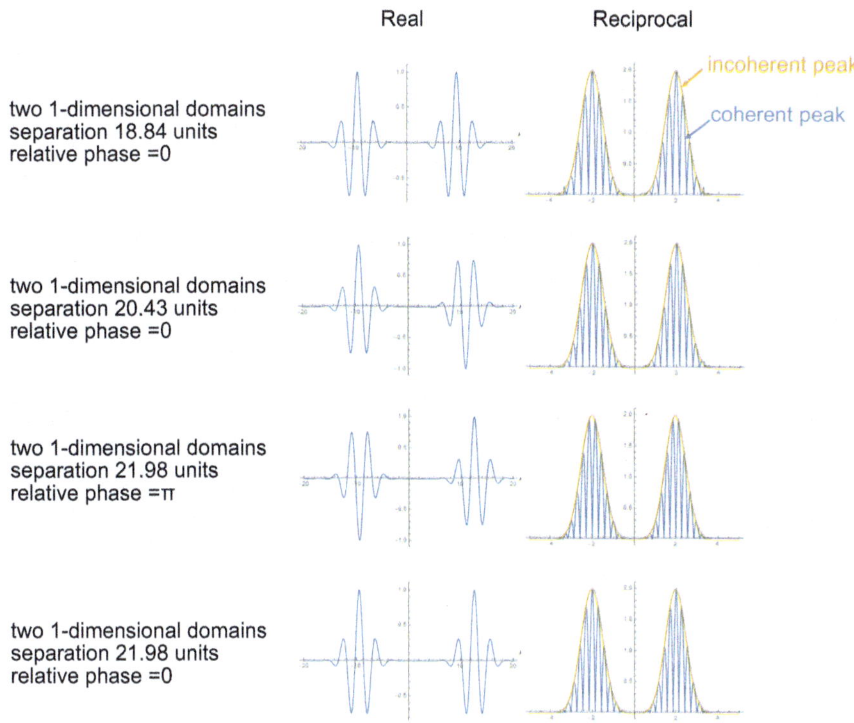

Fig. 2.20 Understanding effect of phase in coherent scattering for 1d: The domains are created from multiplying two Gaussian envelopes in real space with separation and phases as mentioned. The resultant real space and corresponding reciprocal space (by taking amplitude of the Fourier transform) is shown. The separation between two domains increases fringe numbers but the maxima of coherent and incoherent intensity distribution remains aligned. As soon as the phase difference is π for identical separation the relative mainima of the coherent scattering lines up with the incoherent peak position

What about the third dimension? So far we had been solving for the domain distribution in real space, similar to aperture function of a propagating coherent beam. It is only natural to ask why have we ignored the third dimension or the depth of the thin film? This is one advantage of resonant scattering from thin films. The correlation length of the order is as thick as the sample and only the lateral distribution of the domains remains the unsolved question. For single crystal diffraction, if the penetration depth is on the order of the correlation length of the order, this apriori knowledge will remain true. For other cases, such as polycrystalline materials, this condition is not satisfied, and we need solve in 3-dimensions.

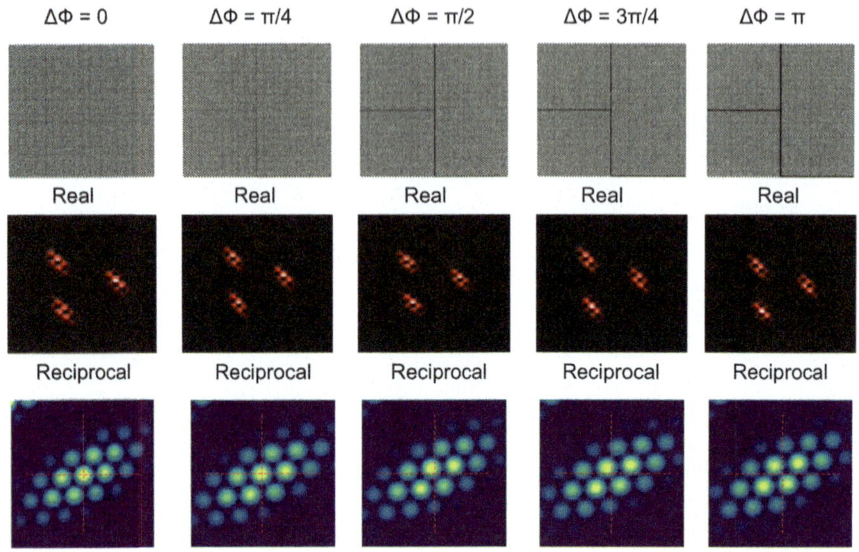

Fig. 2.21 Understanding effect of phase in coherent scattering for 2d: We demostrate here the mechanism of introducing phase difference among the domains by introducing a well defined phase relationship among the underlying textures. When $\Delta\phi = 0$, the coherent scattering intensity distribution hosts maxima of intensity at the location of the incoherent peak maxima (noted by the red dotted cross-hair). As we increase $\Delta\phi$ there is a relative shift between the maxima eventually at $\Delta\phi = \pi$ the minima of the coherent scattering lines up with the maxima of incoherent peak

2.4.3 Detector Orientation in Reciprocal Space

The $\chi = 55°$ scattering geometry used to align the momentum transfer along the pseudocubic body diagonal results in a non-trivial projection of the detector plane onto the H, K, and L reciprocal space directions. This orientation of the detector in reciprocal space is depicted in Fig. 2.22a, b. In order to interpret the small-$\mathbf{q}$ modulations in terms of interfering diffraction from domains separated laterally in the plane of the film ($a - b$ plane), the diffraction pattern is projected onto the $H - K$ plane. This has been done in Fig. 2. After projecting out the L component, the detector direction, having had most L component, is effectively stretched.

In Fig. 2.22d–g we demonstrate the effect of unstretching the real and reciprocal space images after projecting onto the $a - b$ and $H - K$ planes. In the main manuscript, we have chosen to present the reciprocal space patterns and associated real-space domain configurations in their stretched form, thereby emphasizing how the stretching of the $H - K$ plane diffraction pattern onto the skew detector translates into a stretched real-space image following the inverse Fourier transform. The reader should note that it is only possible to project out the L component and directly interpret the modulation patterns on the detector in terms of $H - K$ plane diffraction due to our pre-existing knowledge that the scattering is constrained to the quasi-2D

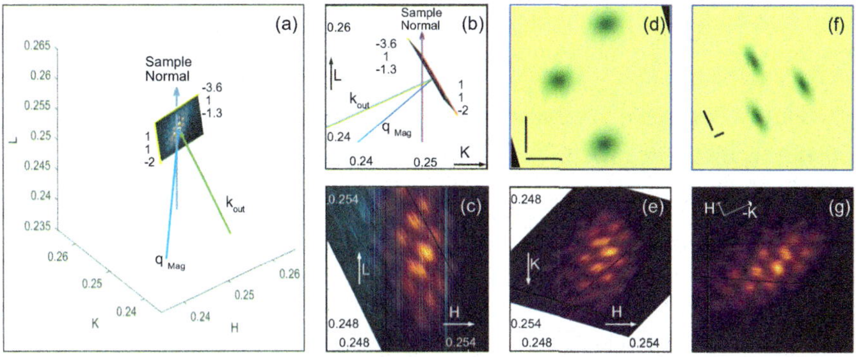

Fig. 2.22 Detector orientation and alternate visualizations of the real and reciprocal space patterns: **a** Depiction of the detector oriented in 3-dimensional reciprocal space. **b** Detector in reciprocal space viewed along the H direction. This perspective captures the strong coupling of K and L components along the detector direction perpendicular to (1 0 0). **g** Experimental diffraction pattern captured by the detector with scale bars in the projected $H-K$ plane (0.001 Å^{-1}) given by the lengths of the white arrows. **c** Same pattern projected onto the $H-L$ plane and **e** projected onto the $H-K$ plane. The peak elongation appears primarily in the L direction. **f** Domain arrangement in the stretched real space depiction whose squared Fourier transform corresponds to the experimentally observed diffraction pattern in (**g**). The scale bars in **f** both correspond to 1 μm along the [1 0 0] (long bar) and [0 1 0] (short bar) directions. The stretching of the real and reciprocal space images in (**f**) and (**g**) results from projecting out the significant L component along one direction of the detector and then interpreting the pattern in terms of $H-K$ plane diffraction. The unstretched real and reciprocal space patterns (equal 1 μm scale bars in both directions) are shown in panels (**d**) and (**e**). In addition to unstretching, the patterns in **d** and **e** have been rotated to align the a/b and H/K directions with the page horizontal and vertical

thin film geometry. In our case, the antiferromagnetic domains being probed are pancake-like objects, with hundreds of nm correlation lengths in the $a-b$ plane, but constrained in the growth direction by the 40 nm film thickness.

2.5 Outlook

In the following outlook section, we consider hypothetical dynamically driven magnetic textures and argue that time-resolved RCXD, in the dilute scattering limit, presents a powerful opportunity to study the motion of antiferromagnetic domains or topological defects in an antiferromagnetic spin texture. Figure 2.23a–e depict a series of hypothetical dilute antiferromagnetic domain configurations in real space. Figure 2.23a corresponds to a single circular domain, whereas Fig. 2.23b–d depict snapshots of one domain passing in between two other domains of equal size, and then disappearing in Fig. 2.23e. The diffraction patterns in the vicinity of the first-order antiferromagnetic Bragg peak originating from each of the real-space domain arrangements shown in Fig. 2.23a–e are simulated in Fig. 2.23f–j respectively. The number and relative positions of the probed domains are clearly reflected in the symmetries of the small-**q** modulations of the antiferromagnetic Bragg peak intensity.

Next we extend the concept of dilute magnetic textures to encompass the notion of a dilute set of topological defects on an antiferromagnetic lattice, such as antiferromagnetic knife edge dislocations observed in the skyrmion nucleation regime of helical antiferromagnets [36], or theoretically predicted antiferromagnetic vortices and skyrmions [37, 38]. Figure 2.23p–t demonstrate the sensitivity of the simulated RCXD cross section around a first-order antiferromagnetic Bragg peak, to the presence and relative arrangements of a dilute set of antiferromagnetic vortices depicted in real space in Fig. 2.23k–o. The diffraction patterns in Fig. 2.23p–t correspond to a resonant magnetic scattering polarization channel which selects the spin x-component. In Fig. 2.23k and p the trivial case of a homogeneously ordered antiferromagnetic lattice is considered, whereas, Fig. 2.23l–m and q–r illustrate the motion of one antiferromagnetic vortex past two stationary vortices. Removing vortices from the beamspot, panels (n) and (s) correspond to just a pair of antiferromagnetic vortices, and panels (o) and (t) to a single vortex.

Sensitivity to macroscopic phase textures: For the purposes of this demonstration, we considered equal antiferromagnetic vortices, each corresponding to a topological defect of winding number $w = 1$, with the spins winding in the plane of the lattice. Similar scattering signatures are expected to arise in the presence of antiferromagnetic skyrmions [37, 38], where the antiferromagnetic spin orientations wind in three-dimensions. Note that in our results, the number of angular nodes in the simulated antiferromagnetic Bragg intensity corresponds to twice the number of $w = 1$ vortices present in the beamspot, or more generally, the net topological charge of the spin texture contained within the beamspot. This indicates an important quantitative sensitivity of the RCXD cross section to the mesoscopic winding of the antiferromagnetic order parameter, reminiscent of the results of Zhang, van der Laan and Hesjedal in the skyrmion material Cu_2OSeO_3 [39], or the resonant x-ray scattering method for determining the topological invariant of nondegenerate band crossings in Weyl semimetals proposed by Kourtis [40].

To model the antiferromagnetic vortices depicted in Fig. 2.23k–o, we assign each lattice site with a complex number, whose real and imaginary parts represent the x- and y-components of its spin orientation

$$m_x = \Re\left(\Pi_n e^{\iota\omega_n \arctan\frac{y-y_n}{x-x_n}}\right) \tag{2.27}$$

$$m_y = \Im\left(\Pi_n e^{\iota\omega_n \arctan\frac{y-y_n}{x-x_n}}\right), \tag{2.28}$$

where (x_n, y_n) and ω_n are the position and winding number of the nth vortex. The product is taken over all vortices in the coherent beamspot. The simulated diffraction patterns shown in Fig. 2.23p–t correspond to the Fourier transform of m_x.

Here we elaborate on the RCXD cross section to macroscopic winding of the antiferromagnetic order parameter. First, we observe existence of a phase texture

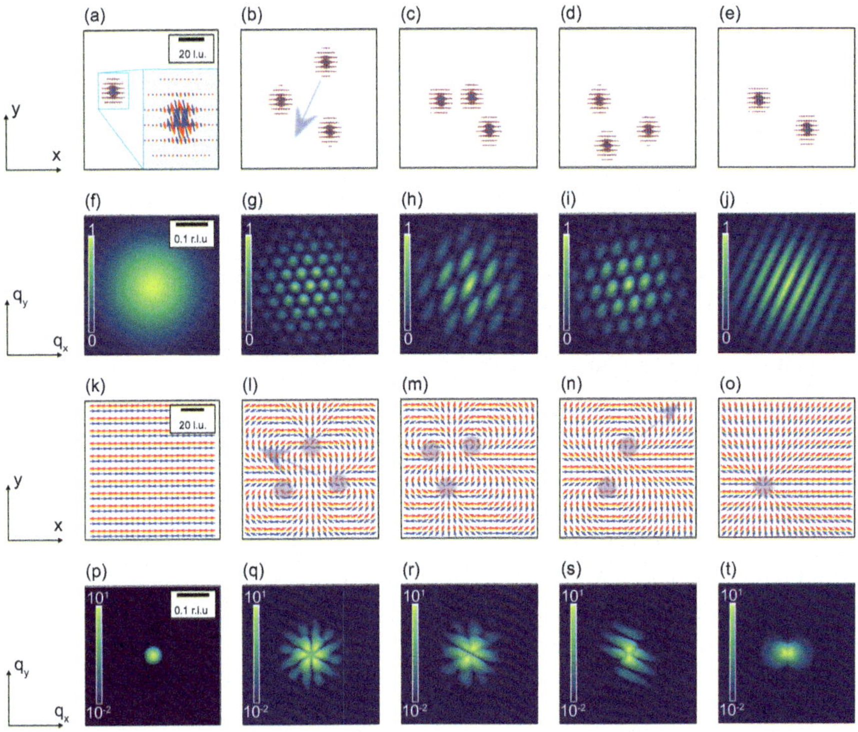

Fig. 2.23 Simulation of RCXD from mobile antiferromagnetic domain arrangements and antiferromagnetic topological spin textures: **a** Single antiferromagnetic domain in two dimensions. **b–d** Three equally-sized antiferromagnetic domains, one moving with respect to the other stationary two. **e** Two equally-sized antiferromagnetic domains. **f–j** Absolute value squared of the Fourier transforms of the model domain configurations in (**a**)–(**e**). **k** Uniform antiferromagnetism in two dimensions. **l–o** Antiferromagnetic vortices with topological winding number $w = 1$. The centers of the vortices are indicated by a shaded disc. **p–t** Absolute value squared of the Fourier transforms of the spin x-component for the antiferromagnetic spin textures shown in (**k**)–(**o**). The number of angular nodes corresponds to twice the sum of the winding numbers of all vortices in the beam spot

under the beamspot is met with a minima at the location of incoherent maxima. This is in-line with observing the relative shift of coherent maxima and incoherent maxima discussed above. Then we note that the number of winding of the texture in real space imprints a signature in the reciprocal space as angular nodes around the central minima of the coherent intensity distribution. The origin is similar to the eigen modes of electron wavefunction of hydrogen atom where the a spherical symmetry of three dimension results into eigen-modes. Here, the two-dimensional circular symmetry results into eigen modes of Bessel functions.

Here we chalk out a simple proof of the above statement: The intensity in reciprocal space (**q**) is given by—

$$I(\mathbf{q}) = |\sum_i D(\mathbf{r_i})e^{\iota \mathbf{q}\cdot\mathbf{r_i}-\Phi_i} F_i(\hbar\omega)(\varepsilon \times \varepsilon' \cdot \hat{\mathbf{m}}_\mathbf{i})|^2, \tag{2.29}$$

Let us consider, for a single domain, the $D(\mathbf{r_i}) = 1$ for all i. The terms have a similar meaning, as previously discussed, except for the new term Φ_i, which encodes the phase difference between individual scatterers.

Abiding with the symmetry of the real space, we can do Jacobi-Anger expansion of the term $e^{\iota \mathbf{q}\cdot\mathbf{r_i}-\Phi_i}$

$$\sum_m i^m \frac{1}{2\pi} J_m(kr)e^{\iota m(\phi-\phi_k)} \tag{2.30}$$

and polar Fourier expansion of $\hat{\mathbf{m}}_\mathbf{i}$ into

$$\int_0^\infty P_{k,m}\sqrt{(}\frac{k}{2\pi})J_m(kr)e^{\iota m\phi}kdk, \tag{2.31}$$

How many times the winding happens dictates the angular symmetry at that k variable yielding a specific eigen-mode with a specific eigen-value of m. Corresponding Bessel function ($J_m(kr)$), accompanying the polar intensity distribution, dictated by the specific m, becomes the primary contributor to the intensity pattern. The following Fig. 2.24 pictorially describes the proof.

2.6 Conclusion

In this chapter, we expanded on the single shot coherent method capable of studying electronic phase heterogeneity. We discussed the background and development of the method from Young's double-slit experiment. Then we studied antiferromagnetic $PrNiO_3$ thin films with coherent resonant scattering. We demonstrated that Bragg coherent diffractive imaging can be extended to study the mesoscopic texture of an antiferromagnetic order parameter using resonant magnetic x-ray scattering. We studied the onset of the antiferromagnet transition in $PrNiO_3$, focusing on a temperature regime in which the antiferromagnetic domains are dilute in the beamspot and the coherent diffraction pattern modulating the antiferromagnetic peak is greatly simplified. We demonstrate that it is possible to extract the arrangements and sizes of these domains from single diffraction patterns and show that the approach could be extended to a time-structured light source to study the motion of dilute domains or the motion of topological defects in an antiferromagnetic spin texture.

In the following chapters we will utilize this powerful method of RCXD to study—

- artificially created antiferromagnetic textures (Chap. 3)
- phase competition of an antiferromagnetic textures (Chap. 4)

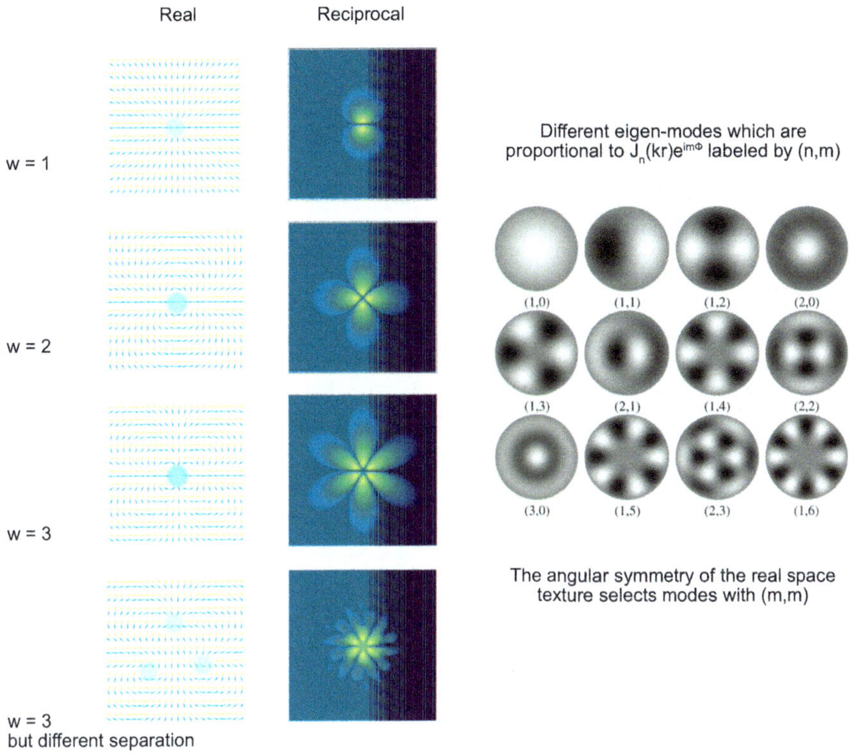

Fig. 2.24 More on studying phase textures with coherent resonant scattering: Simple insight from harmonic analysis and symmetry help us identify character of the winding, their winding number and separation of individual charges. Note the +3 charge and a distribution of three +1 charge gives different scattering intensity, even though the intensities are identical in terms of lobe numbers (twice of winding number). Polar Bessel functions are shown on the right [41]

References

1. Shpyrko OG (2014) X-ray photon correlation spectroscopy. J Synchr Rad 21(5):1057–1064
2. Sinha SK, Jiang Z, Lurio LB (2014) X-ray photon correlation spectroscopy studies of surfaces and thin films. Adv Mat 26(46):7764–7785
3. Shpyrko OG, Isaacs ED, Logan JM, Feng Y, Aeppli G, Jaramillo R, Kim HC, Rosenbaum TF, Zschack P, Sprung M et al (2007) Direct measurement of antiferromagnetic domain fluctuations. Nature 447(7140):68–71
4. Hua N, Li J, Hrkac SB, Barbour A, Hu W, Mazzoli C, Wilkins S, Kukreja R, Fullerton EE, Shpyrko OG (2023) Discerning element and site-specific fluctuations of the charge-orbital order in fe 3 o 4 below the verwey transition. Phys Rev Mat 7(1):014413
5. Song J, Patel SKK, Bhattacharya R, Yang Y, Pandey S, Chen XM, Maple MB, Fullerton EE, Roy S, Mazzoli C et al (2020) Direct measurement of temporal correlations above the spin-glass transition by coherent resonant magnetic x-ray spectroscopy. arXiv:2002.04259
6. Chen XM, Thampy V, Mazzoli C, Barbour AM, Miao H, Gu GD, Cao Y, Tranquada JM, Dean MPM, Wilkins SB (2016) Remarkable stability of charge density wave order in la 1.875 ba 0.125 cuo 4. Phys Rev Lett 117(16):167001

7. Shen L, Esposito V, Burdet NG, Zhu M, Petsch AN, Croft TP, Collins SP, Ren Z, Westermeier F, Sprung M et al (2023) Interplay between relaxational atomic fluctuations and charge density waves in la 2–x sr x cuo 4. Phys Rev B 108(20):L201111
8. Chen XM, Mazzoli C, Cao Y, Thampy V, Barbour AM, Hu W, Lu M, Assefa TA, Miao H, Fabbris G et al (2019) Charge density wave memory in a cuprate superconductor. Nat Commun 10(1):1435
9. Chesnel K, Nelson J, Wilcken B, Kevan SD (2012) Mapping spatial and field dependence of magnetic domain memory by soft X-ray speckle metrology
10. Wochner P, Gutt C, Autenrieth T, Demmer T, Bugaev V, Ortiz AD, Duri A, Zontone F, Grübel G, Dosch H (2009) X-ray cross correlation analysis uncovers hidden local symmetries in disordered matter. Proc Natl Acad Sci 106(28):11511–11514
11. Zaluzhnyy IA, Kurta RP, Scheele M, Schreiber F, Ostrovskii BI, Vartanyants IA (2019) Angular x-ray cross-correlation analysis (axcca): basic concepts and recent applications to soft matter and nanomaterials. Materials 12(21):3464
12. Su R, Seu KA, Parks D, Kan JJ, Fullerton EE, Roy S, Kevan SD (2011) Emergent rotational symmetries in disordered magnetic domain patterns. Phys Rev Lett 107(25):257204
13. Wang Y, Zhou Y, Yang Q, Basak R, Xie Y, Le D, Fuqua AD, Shipley W, Yam Z, Frano A et al (2024) Self-assembly of nanocrystal checkerboard patterns via non-specific interactions. Nat Commun 15(1):3913
14. Scagnoli V, Staub U, Mulders AM, Janousch M, Meijer GI, Hammerl G, Tonnerre JM, Stojic N (2006) Role of magnetic and orbital ordering at the metal-insulator transition in $NdNiO_3$. Phys Rev B 73:100409
15. Boris AV, Matiks Y, Benckiser E, Frano A, Popovich P, Hinkov V, Wochner P, Castro-Colin M, Detemple E, Malik VK, Bernhard C, Prokscha T, Suter A, Salman Z, Morenzoni E, Cristiani G, Habermeier H-U, Keimer B (2011) Dimensionality control of electronic phase transitions in nickel-oxide superlattices. Science 332(6032):937–940
16. Hepting M, Green RJ, Zhong Z, Bluschke M, Suyolcu YE, Macke S, Frano A, Catalano S, Gibert M, Sutarto R, He F, Cristiani G, Logvenov G, Wang Y, van Aken PA, Hansmann P, Le Tacon M, Triscone J-M, Sawatzky GA, Keimer B, Benckiser E (2018) Complex magnetic order in nickelate slabs. Nat Phys 14(11):1097–1102
17. Post KW, McLeod AS, Hepting M, Bluschke M, Wang Y, Cristiani G, Logvenov G, Charnukha A, Ni GX, Radhakrishnan P, Minola M, Pasupathy A, Boris AV, Benckiser E, Dahmen KA, Carlson EW, Keimer B, Basov DN (2018) Coexisting first- and second-order electronic phase transitions in a correlated oxide. Nat Phys 14(10):1056–1061
18. Li J, Pelliciari J, Mazzoli C, Catalano S, Simmons F, Sadowski JT, Levitan A, Gibert M, Carlson E, Triscone J-M, Wilkins S, Comin R (2019) Scale-invariant magnetic textures in the strongly correlated oxide $NdNiO_3$. Nat Commun 10(1)
19. Gibert M, Zubko P, Scherwitzl R, Íñiguez J, Triscone J-M (2012) Exchange bias in $LaNiO_3$-$LaMnO_3$ superlattices. Nat Mater 11:195
20. Frano A, Schierle E, Haverkort MW, Lu Y, Wu M, Blanco-Canosa S, Nwankwo U, Boris AV, Wochner P, Cristiani G, Habermeier HU, Logvenov G, Hinkov V, Benckiser E, Weschke E, Keimer B (2013) Orbital control of noncollinear magnetic order in nickel oxide heterostructures. Phys Rev Lett 111(10):106804
21. Hepting M, Minola M, Frano A, Cristiani G, Logvenov G, Schierle E, Wu M, Bluschke M, Weschke E, Habermeier H-U, Benckiser E, Le Tacon M, Keimer B (2014) Tunable charge and spin order in $PrNiO_3$ thin films and superlattices. Phys Rev Lett 113:227206
22. Bluschke M, Frano A, Schierle E, Minola M, Hepting M, Christiani G, Logvenov G, Weschke E, Benckiser E, Keimer B (2017) Transfer of magnetic order and anisotropy through epitaxial integration of $3d$ and $4f$ spin systems. Phys Rev Lett 118:207203
23. Chakoumakos BC, Schlom DG, Urbanik M, Luine J (1998) Thermal expansion of $LaAlO_3$ and (La, Sr)(Al, Ta)O_3, substrate materials for superconducting thin-film device applications. J Appl Phys 83(4):1979–1982
24. Sutton M, Mochrie SGJ, Greytak T, Nagler SE, Berman LE, Held GA, Stephenson GB (1991) Observation of speckle by diffraction with coherent x-rays. Nature 352(6336):608–610

25. van der Veen F, Pfeiffer F (2004) Coherent x-ray scattering. J Phys: Condens Matter 16(28):5003–5030
26. Eisebitt S, Lüning J, Schlotter WF, Lörgen M, Hellwig O, Eberhardt W, Stöhr J (2004) Lensless imaging of magnetic nanostructures by x-ray spectro-holography. Nature 432(7019):885–888
27. Rodenburg JM, Hurst AC, Cullis AG, Dobson BR, Pfeiffer F, Bunk O, David C, Jefimovs K, Johnson I (2007) Hard-x-ray lensless imaging of extended objects. Phys Rev Lett 98(3):034801
28. Miao J, Ishikawa T, Robinson IK, Murnane MM (2015) Beyond crystallography: diffractive imaging using coherent x-ray light sources. Science 348(6234):530–535
29. Fienup JR (1978) Reconstruction of an object from the modulus of its Fourier transform. Opt Lett 3(1):27
30. Fienup JR (1987) Reconstruction of a complex-valued object from the modulus of its Fourier transform using a support constraint. J Opt Soc Am A 4(1):118–123
31. Bruck YM, Sodin LG (1979) On the ambiguity of the image reconstruction problem. Opt Commun 30(3):304–308
32. Bates RHT (1982) Fourier phase problems are uniquely solvable in more than one dimension: I. Underlying theory. Optik 61:247–262
33. Private communication with Prof. S. Eisebitt
34. Hill JP, McMorrow DF (1996) Resonant exchange scattering: polarization dependence and correlation function. Acta Crystallogr A 52(2):236–244
35. Spence JC, Weierstall U, Howells M (2004) Coherence and sampling requirements for diffractive imaging. Ultramicroscopy 102:149–152
36. Yu XZ, Onose Y, Kanazawa N, Park JH, Han JH, Matsui Y, Nagaosa N, Tokura Y (2010) Real-space observation of a two-dimensional skyrmion crystal. Nature 465(7300):901–904
37. Barker J, Tretiakov OA (2016) Static and dynamical properties of antiferromagnetic skyrmions in the presence of applied current and temperature. Phys Rev Lett 116(14):147203
38. Zhang X, Zhou Y, Ezawa M (2016) Antiferromagnetic skyrmion: stability, creation and manipulation. Sci Rep 6(1)
39. Zhang SL, van der Laan G, Hesjedal T (2017) Direct experimental determination of the topological winding number of skyrmions in Cu_2OSeO_3. Nat Commun 8(1)
40. Kourtis S (2016) Bulk spectroscopic measurement of the topological charge of Weyl nodes with resonant x-rays. Phys Rev B 94(12):125132
41. Wang Q, Ronneberger O, Burkhardt H (2009) Rotational invariance based on fourier analysis in polar and spherical coordinates. IEEE Trans Pattern Anal Mach Intell 31(9):1715–1722

Chapter 3
Artificial Antiferromagnetic Pattern Formation in Nickelate Thin Films

Abstract In this chapter we discuss the fabrication of electronic textures in epitaxial thin films of rare earth nickelates. First, we introduce helium ion implantation as a tool to modify uniaxial strain on an as-grown thin film. We discuss two modalities of helium ion implantation—uniform and focused. We follow it up with a background on nickelates along with a thorough characterization of our $NdNiO_3$ and $PrNiO_3$ thin films. Then, we categorically discuss the effect of uniform helium ion implantation in these thin films leading to a direct observation of localized to itinerant d-orbital. Finally, using spatially sensitive resonant X-ray techniques—Photo-electron Emission Microscopy, Resonant nano-diffraction, and Resonant Coherent X-ray Scattering we observe electronic textures created by local implantation of helium using focused beam. This chapter can be considered as a step-by-step approach to helium ion implantation, beginning with understanding the tool and material selection, moving through implantation effects, and culminating in the fabrication and observation of lateral electronic heterostructures.

3.1 Introduction—Pathway to Pattern Creation

In quantum materials, oxides hold a special spot, with their richness in phenomenology. They host some of the most technologically relevant however least explained phenomena in condensed matter physics such as metal-insulator transition, density waves, high-temperature superconductivity etc. What makes them more exciting is the tunability of the physics they harbor. The ABO_3 perovskite structure is probably one of the most abundant among the different structures where stable states of oxides are formed. A wide combination of A and B site ions is used to form the perovskite unit cell, which distorts from the simple cubic structure to accommodate the differences in internal stress, due to differences in A and B size.

In ABO_3 perovskite phase, a small structural change perturbs the electronic environment strongly to change the response, drastically. We often observe a textured phase diagram with multiple phases competing and co-existing as the structure is tuned. The structural changes from an ideal simple cubic structure can be classified into three main processes: (i) rotations of the BO_6 octahedra (ii) stretching of the

R. Basak, *Nucleation, Fabrication, and Competition in Antiferromagnetic Textures Studied with Coherent Resonant Soft X-ray Scattering*, Springer Theses,
https://doi.org/10.1007/978-3-032-19199-1_3

BO_6 octahedra, and (iii) cation displacement. This is schematically shown in Fig. 3.1 Epitaxial growth of one oxide of interest on top of the motif provided by another perovskite creates further tuning of the overlayer thin film structure [1]. With abundant high-quality substrates available from industrial suppliers, and the advancement of thin film growth technology, epitaxial strain tuning of perovskite structure for tunable functionality has emerged as a very powerful technique. Although there are many stable phases of substrates available for the misfit strain, a quantity capturing the discrepancy of overlayer and substrate lattice mismatch, they only enable tuning in discrete steps. Thus, this method cannot continually tune the overlayer properties.

Furthermore, a fundamental limitation of strain engineering using strain cells or epitaxial growth comes from the Poisson effect, where accommodating a bi-axial in-plane stress in a thin film always involves modification in all three directions. Such modification inhibits a direct experimental way to understand complex physics and further tune it to exact technological needs. This is schematically shown in Fig. 3.2, where an in-plane, bi-axial, compressive strain is accompanied by an out-of-plane tensile strain and vice-versa. In the following section, we will show that, implantation of noble helium ions provides a method to uni-axially tune the strain on thin films from the chemical pressure of intercalated ions. Since ions get implanted into the unit cell, the volume of the unit cell is not preserved anymore.

Finally, as we will see ion implantation provides us with a local tool for strain tuning of the as-grown thin film. Thus, taking advantage of the locality aspect we can create lateral heterostructure of implanted and non-implanted regions. Moreover, differently strained perovskite will undergo structural changes accordingly, resulting in electronic states that differ from each other. Now, if we create a lateral pattern of

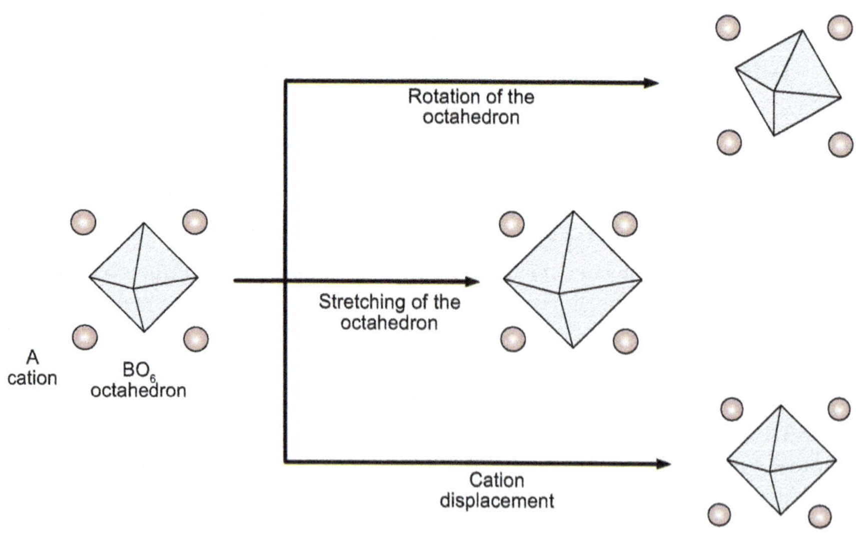

Fig. 3.1 Three primary distortions of ideal simple cubic structures of ABO_3 perovskites: rotation of the octahedron, stretching and shrinking of the octahedron, and cation displacement

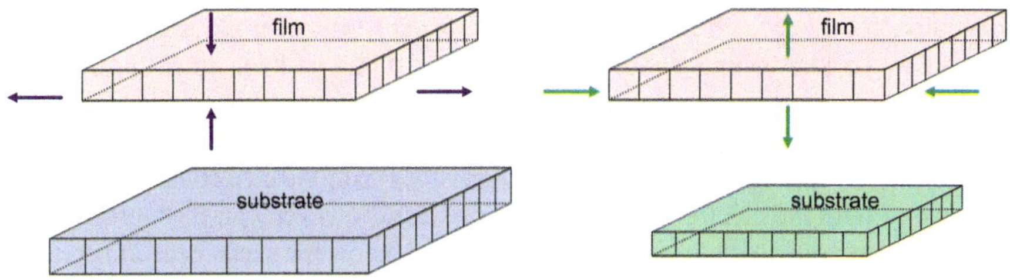

Fig. 3.2 **Poisson Effect in epitaxially grown thin films**: (left) A substrate motif having a larger in-plane lattice constant applies a tensile strain on the overlayer film. As a consequence a compressive strain is experienced by the thin film unit cells along the out-of-plane direction. (right) Similarly a compressive in-plane strain generates an out-of-plane tensile strain

such implanted and not-implanted regions, it may offer us a new paradigm where lateral-interface effects become important. Creating superstructures has non-trivial effects on the electronic bands of the material, especially near the Fermi level which influences the overall property of the material. It is well known, how twisting two graphene layers with respect to each other and stacking them on top of each other, creates a lateral supermodulation, also known as the Moiré pattern. The result is evidenced in the band structure as the weakly correlated material turns into a strongly correlated electron system. With such expectations in mind, we set out to first show that it is possible to create lateral heterostructures of electronic phases in nickelate thin films using a focused helium ion beam. After creating lateral heterostructure we will probe the heterogeneity with Synchrotron-based X-ray techniques which are sensitive to lateral heterostructures of electronic phase, i.e. Resonant Coherent X-ray Diffraction (RCXD).

3.2 Helium Ion Beam as a Tool for Local Structural Tuning

Ion implantation is a time-tested technique in the semiconductor industry, where high-energy ions are accelerated and directed to form a beam before finally being implanted into the target. It is used in the semiconductor industry to induce defects as well as dopants into a crystalline material. As soon as it enters the target it collides with the nucleus and electrons of the host material through electro-magnetic interaction and loses its kinetic energy. With successive collisions, it loses some parts of its energy and finally comes to rest at some depth inside the host material. There are two channels of energy loss:

(i) inelastic collision with electrons, and
(ii) elastic collision with the nucleus.

It requires a greater scope than this thesis to go into the details of high-energy ion kinetics inside the materials. However, we have included a concise discussion in the appendix, complete with references, for interested readers. In this section, we will highlight a few key points about the ion kinetics, while it slows down inside of the material.

Capturing the ion kinetics: The energy loss of incoming high-energy ions inside the material is dominated by the inelastic collisions with the electrons. The inelastic interaction with the electrons is a complex process, however, can be approximately understood as equivalent to a drag force. Thus, just below the surface the ions primarily experience a viscous drag force due to the electrons and lose their energy without much deflection. This continues till the energy drops to a certain limit, where elastic collisions with the nucleus become prevalent. We show here the stopping power (Energy loss per unit length of penetration normalized to the density of the material) of electrons and nuclei of gold for an incoming helium ion beam of energy from the ASTAR calculator of the National Institute of Standards and Technology (NIST) which evaluates stopping power by electrons and nuclei of a material for alpha particles (He^{2+}).

Furthermore, transport of ions in matter (TRIM) simulations successfully captures the kinetics of the ions of given energy inside of a material of given composition and density. In the right panel of Fig. 3.3, we have demonstrated the range of electronic slow-down, and the nuclear slow-down of energetic ions inside of materials. The noticeable feature of nuclear slowing down is evidenced by the deflection of the ion trajectories deep below the surface. Nuclear collisions are a random process that distributes the ions in all possible directions. A small percentage of the incoming ions back-scatter as well after nuclear collision. Thus, a distribution of the implanted ions is seen after all of them come to a stop. The collision with the nucleus creates dislocations of the ions in the material as a result. Such defects are generated inside

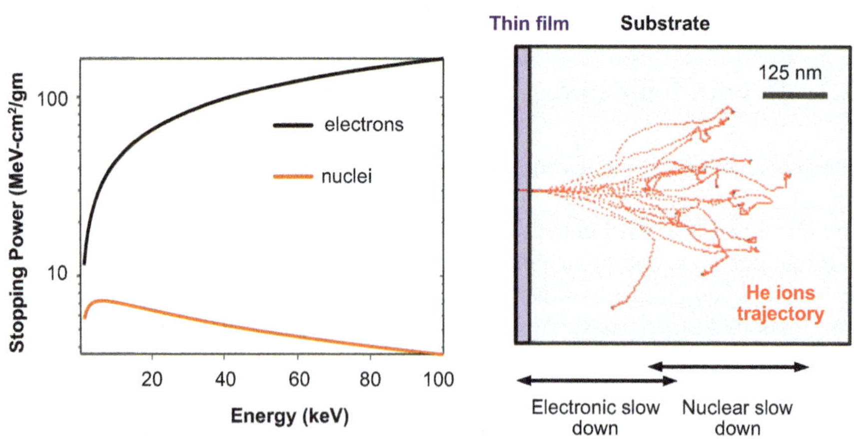

Fig. 3.3 Stopping of helium ions in materials: (left) Helium stopping power of gold electrons and nuclei as a function of helium ion energy (right) Trajectory of 90 keV helium ions in $NdGaO_3$ substrates covered with thin film of $NdNiO_3$, simulated using TRIM simulations

the material with a Gaussian distribution centered at the stopping range of the ion. This is evaluated using the stopping range of ion in matter (SRIM) simulation tool.

3.2.1 *Helium Ion Implantation*

Following our broad discussion of ion implantation of materials we focus now on the helium ion implantation into oxides. Recent experiments have shown that the implantation of noble helium atoms onto epitaxial thin films after growth provides a means to control selectively and continuously the out-of-plane lattice parameters and hence the strain along the out-of-plane direction, while the film remains locked into the bi-axial in-plane strain provided by the substrate [2]. Annealing at high temperatures in the vacuum undoes this uniaxial strain provided by the helium implantation. The authors implanted He ions of 4 keV energy and doses (0, 8, 14, and 21 ions per sq cm) onto 20 nm $La_{0.7}Sr_{0.3}MnO_3$ (LSMO) film epitaxially grown on $SrTiO_3$ substrates. As a result, they showed a continuous expansion of the c-lattice parameter of the film with increasing dose followed by a decrease in the paramagnetic-insulator to ferromagnetic-metal transition temperature, and the magnitude of the magnetoresistance. This work convincingly showed that the continuous tuning of out-of-plane lattice strain provided us with a pathway to continuously tune the electronic properties of strongly correlated electron systems. As highlighted by the authors as well as our previous discussion about the stopping power for the case of lightweight high-energy ions such as helium, the majority of stopping power is contributed by non-nuclear interaction. Thus, they minimize structural damage to the film structure.

The process of He-ion implantation: In the implantation process, He^{2+} ions get accelerated in a static electric field on the order of kV. Using electric fields they are guided towards the grounded film we wish to implant on. The helium current is set corresponding to the dose in ions per square cm (ions/cm^2) we wish to implant the substrate with. Finally, a beam is held at a position for some time (called dwell time) while rastering over a region to match the dose for a given current. Another incredible property of the helium beam is that it can be focused down to 0.5 nm using electromagnetic lenses providing us with the perfect tool to create lateral heterostructures on the thin film. In the Fig. 3.4, we are showing the setup of the helium microscope, which is also used to create the lateral hetero-structures on oxide thin films. Toulouse et al. [3] showed that using such helium microscope (Orion Nanofab), uni-axial strain tuning can be laterally confined in $BiFeO_3$ thin films epitaxially grown on LSAT and STO substrates. Using a focused ion beam they showed that the uniaxial strain tuning was localized into square regions with sizes ranging from 10×10 to 500×500 μm^2.

Once the helium is in the lattice, the stopping of the energetic helium comes from non-nuclear interaction and thus can be modeled using Monte Carlo simulations. Moreover, helium being a noble gas ensures that the sample does not get electron or hole doped. They do not form compounds with the existing ions of the crystal as well. The stability of the implantation exceeds several months without degradation. This has been reported [5], and also tested under the context of the present thesis.

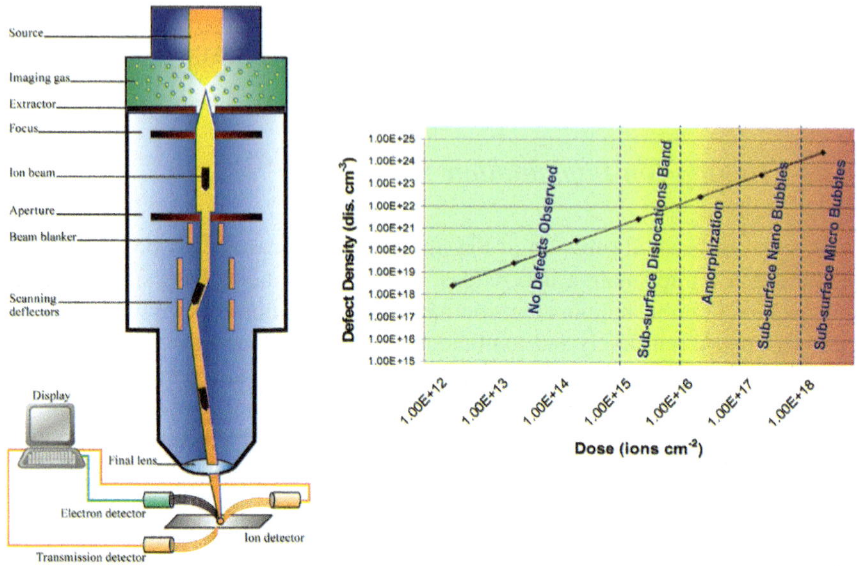

Fig. 3.4 Helium ion microscope and the effect of dose on defect density: (left) The cross-section of the helium ion microscope showing creation, guidance, and focusing of the high energy helium ion beam before implantation on the sample [4]. (right) Defect density observed as a function of dose. The intended dose is to be in the limit where no defects are observed. All of our implantation lies in this regime (below 5e15 ions/cm^2)

The authors studied 20 nm of orthorhombic $SrRuO_3$ coherently grown on $SrTiO_3$ substrates and then exposed to a helium ion beam of 4 keV and several doses. They observed that increasing helium dose directly tuned the out-of-plane lattice parameter of the thin film which in turn tuned the rotation pattern of the BO_6 octahedra and resulted in a more tetragonal unit cell structure from an orthorhombic one.

Defect generation due to He-ion implantation: It is well known in the community that ion bombardment can create knock-on damage to the crystal [4]. This ability of high-energy ion is readily used in focused ion beam facilities to cut off a lamella of materials for microscopic study. However, such damages require high energy of helium ion as shown in the Fig. 3.5, adapted from the topical review [4]. Keeping this plot of defect generation in mind, often the dose for structural strain tuning is kept below 50×10^{14} ions/cm^2, above which sub-surface dislocation bands are observed to be created. Furthermore, in this low-energy and low-dose limit of helium ion implantation functionality of perovskite oxide is seen to be tuned. Hence, we look more into the details about the knock-on damages sustained by the helium implanted perovskite structure.

In this energy limit, the knock-on damages are hard to observe through structural characterization experiments. Thus, we turn to discuss this from a theoretical point of view, following [6]. They used Density Functional Theory (DFT) to evaluate thermodynamic stability point defects that can be caused by helium ion collisions.

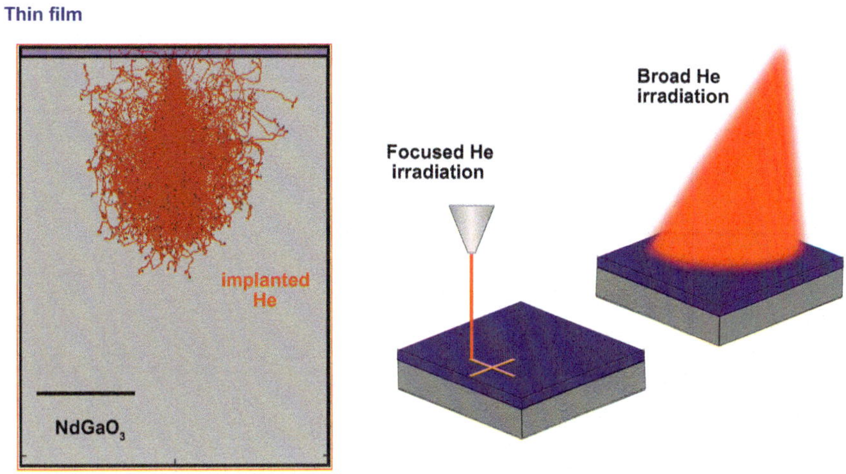

Fig. 3.5 Helium ion implantation in thin films: (left) Cross-sectional view of the thin-film and the substrate along with trajectories of implanted ions until they stop inside. (scale bar: (right) There are two modes of implantation: broad implantation with uniform lateral distribution, or focused implantation with resolution of ~1 nm

They considered their evaluation for 4 keV ions of helium which will be implanted into thin films of $SrRuO_3$ films of 20 nm thickness at a dose of 50×10^{14} ions/cm^2. They considered the helium ions do not propagate into the substrate. Out of three different inherent point defects, (a) vacancy (b) interstitial, and (c) anti-site defects, they found that the lattice parameter expansion along the out-of-plane direction may not have been caused by the He ion interstitial alone. Collisions create cation and oxygen vacancies and the dislodged ions further migrate to interstitial positions. According to the calculation, all of these defects need to be considered as contributing to the lattice expansion.

Setting up the parameter for He ion implantation on the nickelates: For our experiment, we developed thin films of $NdNiO_3$ and $PrNiO_3$, of thickness 15–20 nm on $NdGaO_3$ substrates. We wanted the He ions to go unscattered deep into the substrate, and get implanted. Thus, the thin film only experiences minimum damage, and the most defect formation happens in the substrate of $NdGaO_3$. To facilitate this, we further considered the energy of the ions to be 90 keV. We used the stopping range of ions in matter (SRIM) simulation to evaluate the mean depth of ion implantation for 90 keV ions. For our purposes, we evaluated the numbers such that the helium ions go deep into the substrate and don't stay inside the thin film. The density used for $NdNiO_3$, $PrNiO_3$, and $NdGaO_3$ are 7.24, 7.39, 7.26 gm.cm^{-3} respectively. The figure above shows the ion trajectory simulated using the transport of ions in matter (TRIM) simulation for the following parameters:

- Thickness of the overlayer ($NdNiO_3$) = 20 nm
- Density of the film = 7.24 g.cm^{-3}

Table 3.1 Stopping range of He ions in $NdGaO_3$, $PrNiO_3$, and $NdNiO_3$

Ion energy (keV)	Range in $NdGaO_3$ (nm)	Range in $NdNiO_3$ (nm)	Range in $PrNiO_3$ (nm)
10	46	43	44
25	109	102	105
50	202	188	193
90	327	303	310
100	355	328	337

- Substrate ($NdGaO_3$) thickness = 1 μm
- Density of the substrate = $7.26\,g.cm^{-3}$
- He-beam energy = 90 keV
- incidence angle = 90° *wrt* surface.

In the following Table 3.1, we show the stopping range of He ions of different incident energies. This is the mean depth of the ions from the surface and there is a distribution ∼40 nm at 10 keV, and ∼100 nm at 90 keV. We see that for any energies above 10 keV the ions comfortably cross the thin film of 20 nm thickness and are implanted into the substrates. We wanted to work on this regime so that the helium directly affects the substrate structure, which in turn changes the structure of the thin film through epitaxial relationship.

A hand-wavy way of understanding the effect of helium implantation can be found in analogies to pumping a bike tire. The helium, like the air, goes inside and provides tensile strain only in the direction outward to the surface. For focused ion implantation, we have used 25 keV helium ions, however, for uniform ion implantation we used a higher energy of 90 keV for 20 nm thin films grown on $NdGaO_3$ substrate. We have used the company Kroko inc. for uniform irradiation and used the QB3 facility at UC Berkeley for focused helium ion beam implantation. The energy used for focused ion implantation was limited by the instrument's capability. We have also used samples made by our collaborators of Dynes group at UCSD. These samples were used in the X-PEEM experiment discussed later.

In summary, in this section, we discussed the process of helium ion beam implantation and the incredible uni-axial tuning property it offers. In the following section, we will employ the tuning property to epitaxially grown nickelate thin films and show that lateral heterostructures of electronic states can be created utilizing this tool.

3.3 Phenomenology of Nickelates

In this section, I will first review the physics of rare-earth nickelates relevant in the context of the present thesis. In strongly correlated d-electron systems, such as rare-earth nickelates the central character is that of the nickel d-electrons near the Fermi

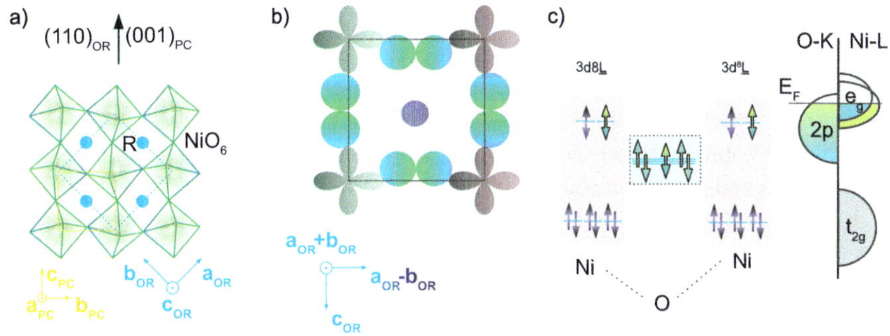

Fig. 3.6 Rich physics of nickelates: a Corner shared octahedra of NiO_6 with the rare-earth ions (R) at the octahedral sites forms the unit cell for $RNiO_3$ **b** Ni-O-Ni bond angle modulates the overlap between $O-2p$ and $Ni-3d$ orbitals schematically described here using orbitals in x-y plane **c** oxygen $2p$ band strongly hybridizes with $Ni-e_g$ orbitals into a covalent $3d^8\underline{L}$ state

level. The short radius of d-electron orbitals makes it the home to strong interactions between electrons that come close to each other. On the other hand, oxygen orbitals are directional and extend from one neighboring d-orbital on one side to the next, allowing pathways for electrons and holes to hop around. These contrasting natures of nickel and oxygen blossom into rich physics in a cubic symmetry such as a perovskite structure where not only the position but the directionality and intermixing of the orbitals play a major role.

In the perovskite rare-earth nickelates, the nickels and the oxygen form a corner-shared network of NiO_6 octahedra as shown in Fig. 3.6a. Although an ideal perovskite maintains cubic symmetry, $RNiO_3$ crystal structure distorts from perfection due to the rare-earth ion size. The smaller size of these ions compared to what is needed for a cubic structure prompts a coordinated distortion in the corner-shared network of octahedra, resulting in an orthorhombic *Pbnm* lattice. Furthermore, from La to Lu along the lanthanides in the periodic table the rare-earth ion becomes heavier and smaller. To accommodate this, the lattice becomes more and more orthorhombic. In Fig. 3.6a, we have shown a typical perovskite structure with the distortion where the rare-earth ions can be used to define the unit cell (marked by purple dashes).

The extent of orthorhombic distortion is measured by the tolerance factor, denoted as t, which is calculated as the ratio of the average separation between rare-earth and oxygen ions (d_{R-O}) to the average separation between nickel and oxygen ions (d_{Ni-O}), $t = \frac{d_{R-O}}{\sqrt{2}d_{Ni-O}}$. Ideally, a perfect cubic structure would yield a tolerance factor of 1. However, what is typically observed in the rare-earth nickelate family, is that this value is often less than 1. Consequently, this leads to a noticeable decrease in the Ni-O-Ni bond angle from its ideal 180°.

Now, when we look closely at the atomic orbital arrangement, adjusting the Ni-O-Ni bond angle away from 180° affects how much the oxygen orbital overlaps with the two nickel orbitals it connects to, thus changing the transfer integral (note Fig. 3.6b). On the other hand, the crystal field environment of the nickel ions splits the otherwise degenerate d orbitals into two sets of orbitals: t_{2g} (those who do not point towards the oxygen ligand orbitals, d_{xy}, d_{yz} and d_{zx}) and e_g (those who point towards the ligand orbitals, $d_{x^2-y^2}$ and d_{3z^2-r2} and raises the e_g orbitals to a higher energy. Furthermore, the filling of orbitals follows Hund's rules of maximum multiplicity: all orbitals should be filled with electrons of the same spin before they are paired. Crystal field and Hund's exchange energy determine the filling up of d-orbitals.

In a nominal ionic configuration of Ni^{3+} the orbital occupation would assume a low-spin state of $t_{2g}^6 e_g^1$. If only the valence d-electrons were the main driver of physics in the nickelates one could integrate out the oxygen orbitals for the renormalized transfer integral between neighboring Ni sites. In such a scenario a partially filled e_g band will have gapless excitation if the constituent electrons are weakly interacting. Instead, if they were strongly interacting with on-site Coulomb repulsion (U_{dd}) the interplay of tunable hopping parameter (t_{dd}) would need to be understood under the light of Mott-Hubbard physics. However, none of these scenarios happen. Neither the nickel ion finds itself in a nominal Ni^{3+} ground state, nor the oxygen orbitals can be integrated out in favor of Mott-Hubbard physics.

Mounting evidence from experiments pointed out that in rare-earth nickelates oxygen 2p band energetically lies between the e_g and t_{2g} bands as shown in Fig. 3.6c and oxygen $2p$ hybridizes with the nickel e_g bands. The energy required for the Since charge transfer energy (Δ_{pd}) dictates the physics of these materials. Experimentally, this is validated by observing the Ni ions in a highly hybridized $3d^8\underline{L}$ state, where $\underline{L}$ denotes a ligand hole [7]. Thus, the ratio of charge transfer energy and the transfer integral becomes a driving parameter according to the Zaneen-Sawatzky-Allen phase diagram for $3d$ Transition metal oxides. Both of these are influenced in a complex way by changing the Ni-O-Ni bond angles.

To summarize, the charge transfer energy (Δ), transfer integral of hopping from $O-2p$ orbital to $Ni-3d$ orbital, on-site Coulomb repulsion, Hund rules, and orbital polarization of the e_g orbital collectively influence the environment of strongly interacting electrons.

What happens if we lower the temperature now?

Phase diagram of rare-earth nickelates: As we lower the temperature, the high-temperature metallic state transitions into a low-temperature insulating state except for R = La, which remains metallic down to the lowest temperature explored. The transition temperature decreases from Lu to Pr consistently with increasing tolerance factor. The insulating phase is found in a bond-ordered and charge-ordered state. The bond-ordered state forms due to a monoclinic ($P2_1/n$) distortion to the orthorhombic ($Pbnm$) high temperature phase. As a result of this distortion, we observe a structural super-lattice formation, where nearest neighbor octahedra shrinks and expands, respectively, from their high temperature metallic state. Furthermore, a concurrent charge order of the same ordering wave vector is observed. On the other hand, an antiferomagnetic order emerges in this bond-distorted insulating state, if we lower the temperature further for $R\mathrm{NiO_3}$ with R = Lu to Sm. The Néel temperature gradually increases with increasing rare-earth size. However, for $NdNiO_3$ and $PrNiO_3$, we observe that both metal to insulator transition and paramagnetic to antiferromagnetic transition are coupled making them the most interesting samples to study the complexity of nickelate physics. Here we study $NdNiO_3$ and $PrNiO_3$ with helium ion implantation.

There are two fundamental attempts motivated from the two different ends of the phase diagram (high and low tolerance factor) to understand the mechanism of the complex phase transition. One of them considers bond order as the primary order parameter and is called the *site selective Mott transition*. Here, the antiferromagnetic order or the charge order is secondary and is not necessary for the phase transition. In this mechanism a short bond makes the oxygen ligand holes to be shared more with the Ni valence shell making it of $3d^8\underline{L}^2$ character forming a singlet. This effectively makes this octahedra detach from the network of octahedra. Since the valence band is completely unoccupied or equivalently, occupied with two electrons and two holes, this site behaves as part of a band insulator. On the other hand, the expanded octahedron gains a $3d^8$ character and the on-site Coulomb repulsion for the half filled valence shell dictates the physics like a Mott insulator. Thus, in this bond-ordered phase two neighboring octahedron sites become insulating following two different mechanisms [8, 9]. However, we need strong electron lattice coupling to explain the charge order formation as a result of the bond order. Recent works have shown that a free energy of the phase with electronic energy f_{el} and electron-lattice coupling is sufficient to explain the proportional charge order emergent in the low temperature nickelate phase [10, 11]. In such a theory, a crucial result is obtained from the free energy analysis which is presented below. The total free energy of the electron-lattice system can be expressed with a harmonic approximation for the lattice energy ($\frac{1}{2}KQ^2$ with Q as the bond order parameter and K is a proportionality constant), along with a coupling term between the electronic charge order, and lattice bond order, and a purely electronic term which encodes a first order phase transition in terms of charge order, ΔN. From symmetry consideration of the order parameter we can write,—

$$f(\Delta N, Q) = \frac{1}{2}KQ^2 - \frac{1}{2}gQ\Delta N + \frac{1}{2}\chi_0^{-1}\Delta N^2 + \frac{1}{4}\beta\Delta N^4 + \frac{1}{6}\gamma\Delta N^6, \quad (3.1)$$

We have incorporated different proportionality constants as they required. The stationary solution with respect Q gives us an important result:

$$KQ = \frac{1}{2}g\Delta N \quad (3.2)$$

We will test this relationship as a measure of the electron lattice coupling, and how it differs from sample to sample. We consider this picture in the nickelates as a local electron picture where the bond order can influence the electrons locally and modify the electronic correlations. The other limit of this is explained by a highly itinerant electron system where they form Fermi surface. For high tolerance factor, this limit is believed to have been experimentally realized [12, 13]. In this limit, a Fermi surface instability generates a spin density wave at $\mathbf{q} = (1/2, 0, 1/2)$ wave vector [14, 15]. We will see during our exploration in nickelates that we are able to change the itinerant versus localized nature of the d-electrons, which in turn changes the character of the magnetism from a localized *site-selective Mott* picture to possibly a *spin density wave* picture. For now, we focus on the two specific samples of $PrNiO_3$ and $NdNiO_3$, which act as a reference sample quality. We will implant helium on them to change their electronic character.

3.4 Results: Coupled Phase Transitions in $NdNiO_3$ and $PrNiO_3$

In this section, we present our experimental exploration of the phase transitions in $NdNiO_3$ and $PrNiO_3$, focusing on the characterization of the ordered phase at low temperatures. To facilitate our discussion, we provide a concise representation of the phase diagram, highlighting the phases under examination. Our aim here is to introduce the phase transitions observed in epitaxially grown rare-earth nickelate thin films of $NdNiO_3$ and $PrNiO_3$ on (110)-oriented $NdGaO_3$ substrates, followed by interpreting the observed data, in the light of site-selective Mott transition picture with strong electron-lattice coupling.

3.4.1 Charge Disproportionation Driven by Bond Order

Before going into the details of the charge order phase transition, let us first introduce the samples and discuss their characteristics. We will discuss their structural details focusing on how the lattices respond to the strain environment. Then we will establish the high quality of our films being studied from the sharpness of the metal-insulator transition we observe. Finally, we will extract charge order and bond order from these samples as a function of temperature.

Structural Characterization: In our discussions, we primarily utilize epitaxial thin films grown via pulsed laser deposition. For comprehensive information regarding thin film growth, please consult the appendix. We have uniformly produced thin films with a thickness of approximately 22 nm on 110-oriented $NdGaO_3$ substrates for both $NdNiO_3$ and $PrNiO_3$ films. Their structures were characterized using an in-house $four - circle$ diffractometer (make: Huber) operating with Cu-K_α source at room-temperature. The specular reflection of (2 2 0) and the off-specular reflection of (2 2 1) were investigated to extract the orthorhombic lattice parameters as shown in the Table 3.2.

Ideally, one needs at least three mutually orthogonal peaks to refine the six parameters (a, b, c and α, β, γ) that define the structure of a 3-dimensional lattice. However, we get around that by the following crucial assumptions based on the works of Valilionis et al.

Table 3.2 Bulk lattice parameters of $RNiO_3$ (R = Pr, Nd) compared with measured lattice parameters of $NdGaO_3$ substrate used. In-plane strain value calculated from them is tabulated $\parallel$ or $\perp$ to c direction of $NdGaO_3$ substrate. We used the formula $100 \times (a_{film} - a_{substrate})/a_{substrate} = \%strain$

Attributes	$PrNiO_3$ bulk	$NdNiO_3$ bulk	$NdGaO_3$ substrate
a (Å)	5.42	5.39	5.37
b (Å)	5.38	5.38	5.56
c (Å)	7.63	7.61	7.73
$\sqrt{a^2+b^2}$ (Å)	7.64	7.62	7.72
In-plane strain $\parallel$ c	1.31	−1.55	1
In-plane strain $\perp$ c	1.04	−1.48	1

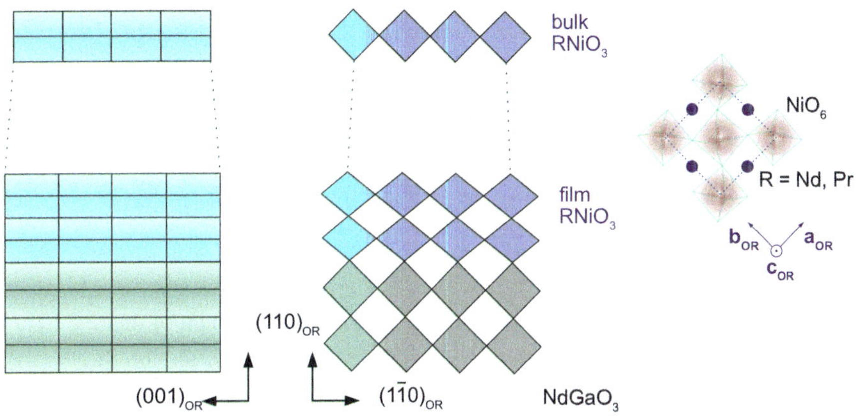

Fig. 3.7 **Effect of misfit strain from substrate epitaxy**: Epitaxy changes the lattice parameters a, b, c and γ. The unit cell is shown for continuity

- the misfit stress between the substrate and the film is accommodated along the (0 0 1) direction differently than $(1\bar{1}0)$ direction.
- the volume of the unit cell is not preserved to accommodate strain

As a consequence, we need to look at the distortion of the octahedra along the two perpendicular in-plane directions separately. Along the (0 0 1) direction, as shown in the Fig. 3.7, the stress is accommodated by elongating the c lattice parameter of the film. The angles α and β remain equal to 90°. Both of these points are confirmed by checking the epitaxy of the film on the substrate off-specular peak of (2 2 1). On the other hand, in the plane spanned by (1 1 0) and $(1\bar{1}0)$ the angle γ is free to change along with the lattice parameters (a, b) and thus, the stress from the substrate is going to be accommodated by all: a, b and γ. Hence, with the condition of epitaxy, a, b and γ are the only unknowns.

Table 3.3 Bulk lattice parameters of $RNiO_3$ (R = Pr, Nd) compared with measured lattice parameters of $NdGaO_3$ substrate used. In-plane strain value calculated from them is tabulated ∥ or ⊥ to c direction of $NdGaO_3$ substrate

Attributes	$PrNiO_3$ film	$NdNiO_3$ film	$NdGaO_3$ substrate
a (Å)	5.42	5.4	5.37
b (Å)	5.42	5.4	5.56
c (Å)	7.73	7.73	7.73
$\alpha(°)$	90	90	90
$\beta(°)$	90	90	90
$\gamma(°)$	90.97	91.33	90

The problem is further simplified by observing the results discussed in where under tensile strain, the a and b lattice parameters of the film, become equal, $a_{film} = b_{film}$. So, finally, we have two unknowns a_{film} and γ_{film} in the two equations as follows:

$$\begin{aligned} \sqrt{a_{film}^2 + b_{film}^2 - 2a_{film}b_{film}cos(\gamma_{film})} &= d_{1\bar{1}0 \rightarrow substrate}, \\ \sqrt{a_{film}^2 + b_{film}^2 + 2a_{film}b_{film}cos(\gamma_{film})} &= 4 \times d_{220 \rightarrow film}. \end{aligned} \tag{3.3}$$

Based on the arguments presented above and our measurement of the (2 2 0) and (2 2 1) peaks we find the following lattice parameters for the film and the substrate (Table 3.3).

The condition of epitaxy along (0 0 1) is confirmed by doing an HKL-scan along (1 1 0) direction (parallel to the growth direction) after aligning to the substrate (2 2 1) peak and finding the film peak at the correct (H, K) coordinates.

Transport Characterization: Four-point van der Pauw measurements were performed to characterize the metal-insulator transition. All of our films discussed in this thesis showed sharp metal-insulator transition with hysteresis loop, as expected from $NdNiO_3$ and $PrNiO_3$ samples. This indicates the high quality of coherent thin film growth.

Resonant Hard X-ray Scattering: Now that we have established the quality of the film, and characterized the structure and metal-insulator transition we move to the central investigation regarding charge order. Hard X-ray resonant scattering experiments were carried out to reveal the charge and bond order of the neighboring NiO_6 octahedra as the sample was cooled ($Pbnm$ to $P2_1/n$). The experiments were carried out in the 6-ID-B beamline of the Advanced Photon Source of Argonne National Lab using a six-circle diffractometer. We have seen previously that a Landau theory of strong electron-lattice coupling produced a direct proportionality between the order parameter strength of the bond order and charge order. To test that prediction, we set out to measure the relative strength of bond order and charge order as we lower the temperature. As shown in Fig. 3.8, we consider the deviation from the average Ni-O

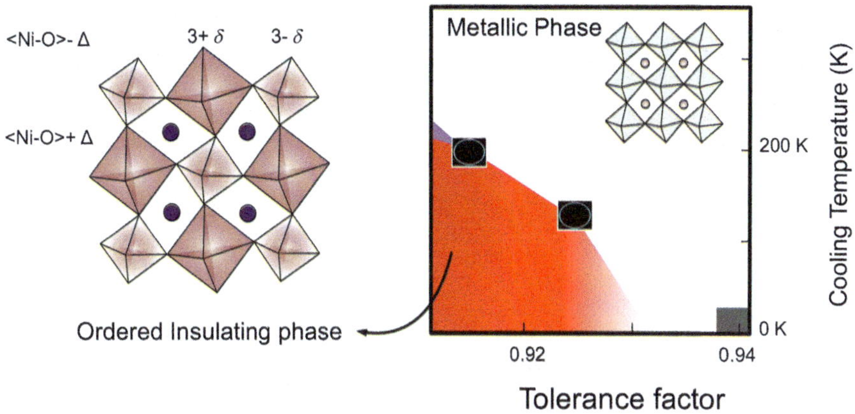

Fig. 3.8 Describing bond and charge order: Bond order (Δ) and charge order (δ) parameters in the relevant section of the phase diagram for the discussion of this section

bond length in each octahedron as the measure of the bond order parameter (Δ). Similarly, the deviation from the average value of charge for neighboring octahedra is considered as the charge order parameter (δ), shown in the Fig. 3.8 for a nominal charge on Ni ion [16].

Figure 3.9 shows the schematic experimental setup for the hard X-ray scattering experiment. The incoming polarization was set to σ to increase scattering sensitivity to charge. The scattered beam was not analyzed. We performed fixed-q energy scans around Ni - K edge with an energy resolution of 1eV (comparable to the broadening by core-hole lifetime) while the diffractometer was oriented to the (011) peak. A fixed-q scan updates the diffractometer position for every energy step for the scan. Since the peak is forbidden without charge asymmetry in nearest neighbor octahedra in the $Pbnm$ phase, we chose this for maximum contrast during phase transition. However, in reality, some monoclinic distortion gives rise to observing this orthorhombic symmetry forbidden peak in the metallic state too but with much lower intensity. Lowering the temperature in these thin films finds a bond and charge-ordered insulating state with a doubling of the unit cell (Fig. 3.8), thereby allowing the increased intensity of this peak. We measured energy dependence of the intensity of the peak as we cycled the temperature of the sample stage using a Liquid He cryostat with a temperature resolution of ± 0.1 K. For each temperature, after necessary centering and realignment due to thermal drift, we varied the energy of the monochromator and measured the beam current normalized intensity of the peak. The order parameters of the bond order and charge order can be extracted from the energy dependence of this scattering intensity.

Let us focus on the extraction of the relative strength of bond order and charge order from the hard X-ray Resonant Elastic Scattering experiments. Raw data from the energy scans across the Ni-K edge on (0 1 1) reflection for both $NdNiO_3$ and $PrNiO_3$ are shown in Fig. 3.10 respectively. Only the data for warming is shown

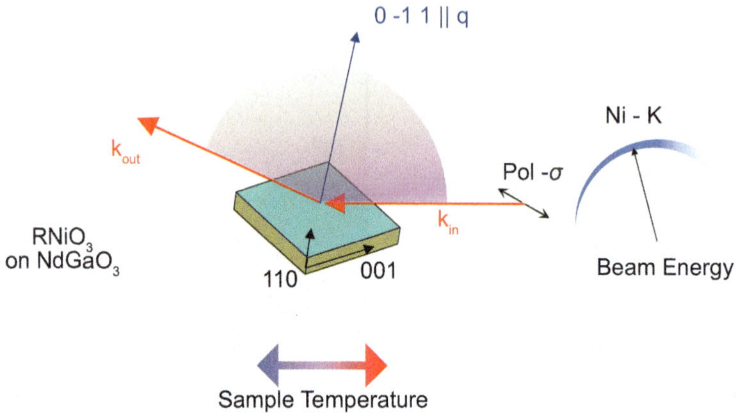

Fig. 3.9 Experimental set up: Experimental set-up for charge order and bond order determination with vertical scattering geometry at APS 6-ID

for clarity. K edge excitation creates a core hole in the 1s state and in the dipole approximation, emission lines probe the unoccupied 4p state when that core-hole collapses. Even though the level states are far above the Fermi level, electronic ordering influences the energies of these states in a complex way. Fortunately, this is picked up simply by the 4p density matrix which gets off-diagonal terms due to the monoclinic distortion in a cubic eigenbasis. The scattered intensity observed, $I(\mathbf{q}, E) = |F(\mathbf{q}, E)|^2$, where $F(\mathbf{q}, E)$ is the structure factor and we have neglected the Debye-Waller correction. The structure factor can be expanded as

$$F(\mathbf{q}, E) = \sum_i e^{\iota \mathbf{q} \cdot r_i} f_i(\mathbf{q}, E)$$

where i runs over all atoms of the orthorhombic or monoclinic unit cell. $f_i(q, E)$ contains energy independent non-resonant$(f_i^n r(q))$ scattering and energy-dependent resonant contribution $(f_i^r(E))$. The energy-dependent scattering factor is a complex number and comprises real and imaginary parts as seen previously. The imaginary part of the scattering factor is directly linked to the X-ray absorption spectroscopy (XAS), whereas the real and imaginary parts both influence the scattering cross-section. Since the imaginary component of the non-resonant scattering factor is very small compared to the real component without loss of generality, we consider $f_i^n r(q)$ to be real. Then, the scattering factor expands into its full form as follows:

$$F(\mathbf{q}, E) = \sum_i e^{\iota \mathbf{q} \cdot r_i} [f_i^{nr}(q) + Re[f_i^r(E)] + \iota Im[f_i^r(E)]]$$

Now as we sum over the unit cell members every term collects phases. The phases due to translation are accounted for by the phase factor $e^{\iota \mathbf{q} \cdot r_i}$, however, the

Ni ions are not only related by translational symmetry. A $(h0l)$ or $(0kl)$ peak with $h, k, l \in Odd$ is forbidden by the translational symmetry of a $Pbnm$ lattice. Thus, if we look at the intensity at Ni resonance there should be no contribution due to the resonant contribution. On the other hand, oxygen ions(O) and the rare-earth ions (R) contribute to the non-resonant part due to monoclinic distortion present as the cooperative bending of the octahedra even in the $Pbnm$ phase.

Let us look closer at the resonant contribution.

In general, $f^r(E) = \sum_{\alpha,\beta=x}^{z} \varepsilon_{o,\alpha}^* \mathcal{F}_{\alpha\beta}\varepsilon_{i,\beta}$, where $\mathcal{F}_{\alpha\beta}$ is the scattering tensor expanded as follows:

$$\mathcal{F}_{\alpha\beta} = \sum_{m=x}^{z} \frac{\left\langle s|\hat{T}_\beta|p_m\right\rangle\left\langle p_m|\hat{T}_\alpha|s\right\rangle}{E - \delta E + \iota\Gamma/2} \tag{3.4}$$

This denotes that the dipole operator $\hat{T} = (\hat{T}_x, \hat{T}_y, \hat{T}_z)$ acts on the spherical 1s orbital and creates a core-hole pair in the respective 4p orbital. In the monoclinic phase, two neighboring octahedra become inequivalent through bond order and charge order which interferes with the density matrix for the unoccupied 4p shells (to be occupied once we create the core-hole). Thus, in the most general consideration, they will contribute a phase factor to the $f^r(E)$ between two neighboring Ni atoms. Since we know the order creates a $2 \times 2 \times 2$ supercell the only phase factor allowed by this symmetry is $e^{\iota\pi} = -1$. Thus, for a unit cell in orthorhombic or monoclinic symmetry, the structure factor for a peak such as the one we investigate (011) becomes:

$$F_{011}(\mathbf{q}, E) = A_{RO}(\mathbf{q}) + \Delta Re[f_{Ni}^r(E)] + \iota\Delta Im[f_{Ni}^r(E)],$$

where $A_{RO}(q)$ contains all the non-resonant contribution from the rare-earth and the oxygen ions.

Finally, when we evaluate the intensity of the peak due to complex conjugation, all contributions from $Im[f_{Ni}^r(E)]$ cancel out leaving only $|Im[f_{Ni}^r(E)|^2$ term [17]. On the other hand, the real part survives with the magnitude and the cross terms with the non-resonant part. Assuming $|A_{RO}(q)| >> |Re[f_{Ni}^{r'}(E)]|, |Im[f_{Ni}^r(E)|$, we can simplify the expression to the following:

$$I_{011}(\mathbf{q}, E) \approx |A_{RO}(\mathbf{q})|^2 + A_{RO}(\mathbf{q})\Delta Re[f_{Ni}^r(E)]$$

Main Result: $\Delta f_{Ni}^r(E)$ is extracted for our case from the raw data using $E_{N Res}$ = 8356 eV (black arrow) as the non-resonant energy and shown in Fig. 3.10 for $NdNiO_3$ and $PrNiO_3$ respectively. The first column shows the raw intensity as measured during the experiment. The second column corresponds to the extracted $\Delta Re[f_{Ni}^r(E)]$. We then note that at the resonance peak (noted by the green arrow) the intensity is mostly dominated by the Ni electronic states. However, at off-resonance (blue arrow), majorly crystal structure determines the intensity of the peak. Thus, we extract the bond order parameter from the raw intensity counts at the non-resonance position below the resonance peak(E_{NR}). It is interesting to note that even the $\Delta[f_{Ni}^{r'}](E)$

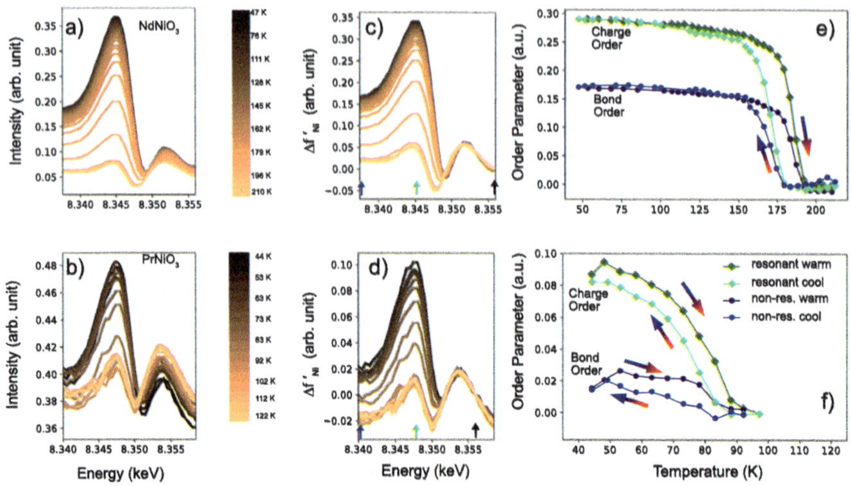

Fig. 3.10 Bond order and charge order evolution with temperature: The top row is for $NdNiO_3$ and the bottom row is for $PrNiO_3$. The left-most columns show raw data. The middle columns show extracted $\Delta f'_{Ni}$ whereas the right-most column shows the extracted bond order and charge order for both warming and cooling cycle

below the resonance peak is significantly influenced by the bond order. However, above the resonance energy, it does not.

Thus, an estimate for the charge order can be extracted from plotting $\Delta f'_{Ni}(E_{Res})$ as a function of temperature while subtracting the high-temperature value. Similarly, we extracted the bond order strength using the temperature evolution of the $\sqrt{I(E_{NR})}$. This is shown for both the warming and cooling cycles as indicated by the color-graded arrows in the last column of Fig. 3.10.

Discussion: From the plots of Fig. 3.10e–f it can be seen that the bond order and charge order in the $NdNiO_3$ film appear simultaneously as expected from the strong electron-lattice coupling. In line with the expectation, the phase transition is sharply first-order in nature with a hysteresis loop of 15 K. We contrast this temperature dependence with the temperature dependence of $PrNiO_3$ where a smaller hysteresis loop, of about 5 K, opens and the orders set in more gradually, indicating a weaker first-order nature of the coupled phase transition. Furthermore, the strength of the order parameters is much lower for $PrNiO_3$ compared to $NdNiO_3$.

Before discussing the comparison, it is worth highlighting a few points here. The order parameters are extracted relative to their high-temperature value. Thus, effects of background, difference in incoming flux, and detector sensitivity can be ignored. Although we kept incoming flux and detector sensitivity identical for both samples. Furthermore, we used the same non-resonant energy to extract the resonant energy evolution of $\Delta[f^{r'}_{Ni}]$ to have them comparable to each other. In other words, both the orders are expressed in the same units even though the unit is not Å for the bond order or fraction of electron charge for the charge order. Hence, we have plotted the charge order strength with the bond order strength of the respective systems in Fig. 3.11.

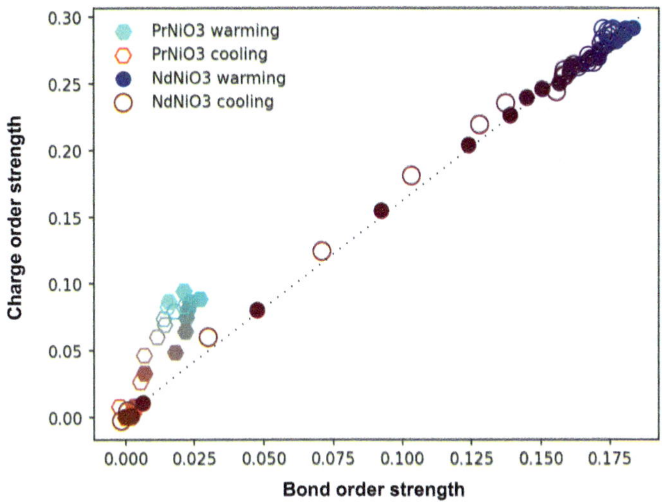

Fig. 3.11 Bond order and charge order parameter co-dependence: For both $NdNiO_3$ and $PrNiO_3$ we observe a linear relationship between the bond order and the charge order extracted

We first note the linear relationship observed by the bond and charge order indicating strong electron-lattice coupling as suggested by and observed by in $NdNiO_3$-NAO superlattice. Both $NdNiO_3$ and $PrNiO_3$ show the linear dependence, although the proportionality constant is slightly larger for $PrNiO_3$. In the view of the model laid out by [18], this could mean a lower coupling between electron and lattice. For a more itinerant d-electron system of $PrNiO_3$, it is expected to have lower electron-lattice coupling. However, our results indicate that even though it is weaker in amplitude than $NdNiO_3$ charge order, it still follows a linear relationship with the bond order.

Secondly, we have extracted the critical exponent of the phase transition shown in Fig. 3.12. We did a time scan, i.e. we parked the detector on the (011) peak and let the temperature rise. We made sure that our region of interest in the area detector captured the peak. The thin films were large enough to ensure that the cryostat expansion during this measurement did not alter the sample-beam intersection. For a first-order phase transition, the critical exponent can be related to the percolation of the nucleating phase. We subtracted the high-temperature intensity ($Intensity(T_{High})$, where T_{High} is well above the transition temperature) from the raw counts and then fit the result to the expression $(1 - \frac{T}{T_C})^{2\beta}$ to extract both T_c and β. The β also differ significantly between $NdNiO_3$ and $PrNiO_3$. This may indicate that the phase texture evolution in $NdniO_3$ is different than $PrNiO_3$.

Thus, from this discussion, we conclude that $NdNiO_3$ hosts a stronger electron-lattice coupling and thus electronic behavior can be tuned if we can tune the lattice. Thus, $NdNiO_3$ is an immediate choice for strain tuning via helium implantation. In the case of $PrNiO_3$, this is complicated and will be harder to control since a small bond order change resulted in a relatively stronger charge order change. This may indicate

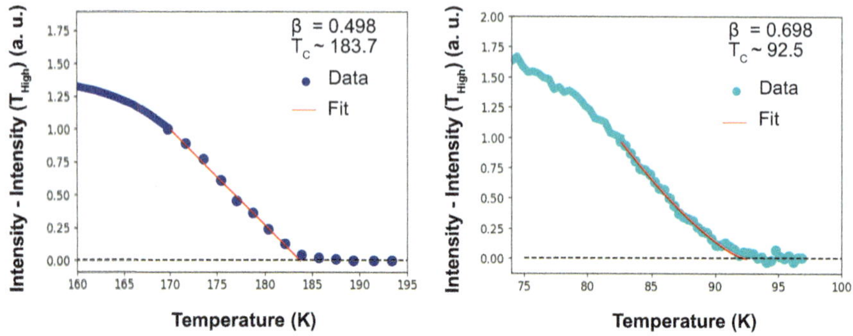

Fig. 3.12 Critical exponent of the phase transition: (left) $NdNiO_3$ (right) $PrNiO_3$

other mechanisms such as a Pierels instability arising from spin density wave in an orthogonal symmetry are at play. We speculate that such a mechanism is on top of the site-selective Mott transition causing the increased charge order observation. With this insight, we move to the next section where we highlight soft X-ray scattering investigations done on these samples.

3.4.2 *Antiferromagnetism*

Both of our $NdNiO_3$ and $PrNiO_3$ films transition into a low-temperature non-colinear antiferromagnetic phase with an ordering vector of $(\frac{1}{2}, 0, \frac{1}{2})$, consistent with the phase diagram (Fig. 3.13). To characterize the magnetic phase and subsequently compare the two magnetic phases between $NdNiO_3$ and $PrNiO_3$ we conducted soft X-ray resonant scattering. In this case, soft X-ray energy was tuned to the Ni-L3 edge. Soft X-ray magnetic scattering was carried out in a Kappa-diffractometer of 29-ID-C beamline of Advanced Photon Source of Argonne National Lab. Kappa-diffractometer can be directly mapped into a four-circle diffractometer with constraints. On these thin films, we have also performed soft X-ray absorption spectroscopy in total electron yield mode for Ni-L3 and O-K edges towards a complementary understanding of the electronic structure in the beamline of 4.0.2 in Advanced Light Source of Lawrence Berkeley National Lab.

An antiferromagnetic order such as the one observed in nickelates can be characterized by (i) the temperature dependence of the magnetic order parameter, m (ii) the correlation length of the order (iii) the energy dependence of the magnetic order, and (iv) its magnetic moment orientation. Here we have explored the first three for all of our films. Before we discuss the results, let us first go through the experiment methodology.

This begins by finding the in-plane orientation of the substrate. First, we figured out the (001) direction from finding the (221) peak. Usually one of the edges of the

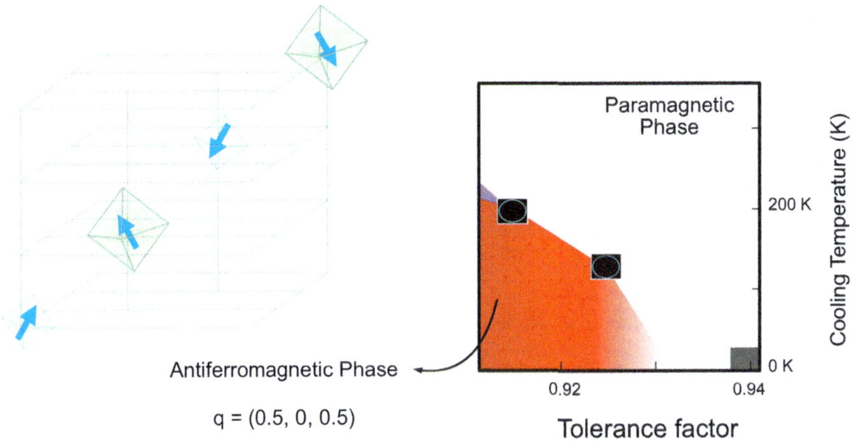

Fig. 3.13 Describing the antiferromagnetic order: Schematic of the phase diagram and the low-temperature antiferromagnetic phase with a spiral structure

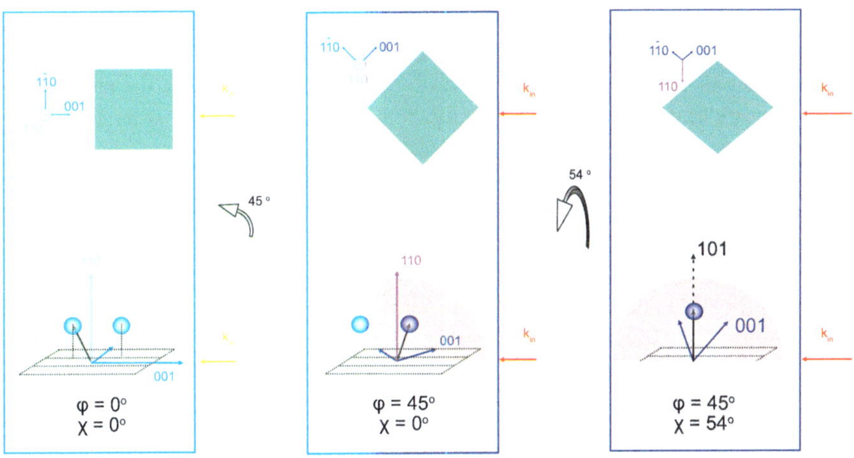

Fig. 3.14 Experimental method to do the resonant scattering experiment: each step from left to right shows how the sample is rotated in real space (top row) and as a result how the peaks in the reciprocal space get oriented with respect to the scattering plane. After the last column we change 2th to 110° and theta to 55° to observe the magnetic peak

as-bought substrate is parallel to the (001) direction. Then, we remounted the thin film in the diffractometer with ($1\bar{1}0$ in the scattering plane. Next, we resolved ($\bar{1}10$) from $1\bar{1}0$ by finding the (240) and (420) respectively. Now that we have confirmed the orientation we took the samples to the Kappa diffractometer in 29-ID-C. Finally, we mounted them with (001) pointed towards the beam at th = 0°, and sample normal pointed towards 2th = 90° detector position. At low temperature and Ni-L3 resonance with phi rotated by 45° and chi by 54° we could see the magnetic peak

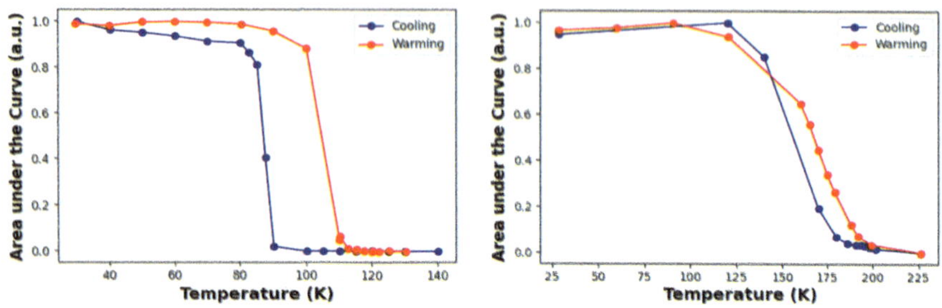

Fig. 3.15 **Antiferromagnetic order evolution with temperature**: (left) $PrNiO_3$ and (right) $NdNiO_3$: Both transition shows a first order nature with clear hysteresis loop

on our detector. This is schematically shown in Fig. 3.14. Note, after the ϕ rotation a $-54°$ of χ rotation would not have brought any one of the magnetic peaks into the scattering plane. The Kappa diffractometer had chi and phi-motor equivalent motion, otherwise, the standard practice is to make a wedge of 54° and mount the sample phi-rotated by 45°. This way at the resonance condition the magnetic peak is found in symmetric mode at 2th = 110°. We should mention that apart from this symmetric mode there can be many asymmetric ways ($th \neq 2th/2$) of finding the peak. However, we stick to this mode for simplicity from the geometrical consideration of the Kappa diffractometer (Fig. 3.15).

Result and Discussion: Once we found the magnetic peak, we measured the temperature dependence by taking rocking curve scans after aligning to the peak at each temperature. We used a Si diode detector with an angular span of 3°. The response was measured by an Ammeter connected to the diode. For every scan, we evaluated the area under the curve after subtracting the background and fitting the signal with a Gaussian peak. The result of the measurement is shown in Fig. 3.16. Consistent with the metal-insulator transition and bond order-charge order transition the coupled antiferromagnetic transition also occurs around 90 K for $PrNiO_3$ and around 210 K for $NdNiO_3$. The correlation length for $NdNiO_3$ is significantly larger than the correlation length of $PrNiO_3$ indicating a more rigid antiferromagnetic phase. The correlation length was observed to be 45 nm for $NdNiO_3$ and 29 nm for $PrNiO_3$. The observed antiferromagnetic correlation length is similar to studies conducted elsewhere.

Energy response of the magnetic peak: Next, we turn our attention to the spectroscopic information that we harnessed from the samples. First, let's discuss the energy dependence of the magnetic peak. Similar to the measurement done at the Ni-K edge, here we also varied the energy while the diffractometer was set at the momentum transfer condition $\mathbf{q} = (\frac{1}{2}, 0, \frac{1}{2})$. In other words, for every energy we re-oriented our diffractometer to match the diffraction condition for the magnetic Bragg peak for the given energy. The energy dependence revealed a clear Ni-L3 resonance as shown. Moreover, on closer inspection, we see that the energy dependence shows a two-peak structure. We fit the energy dependence with a large Lorentzian

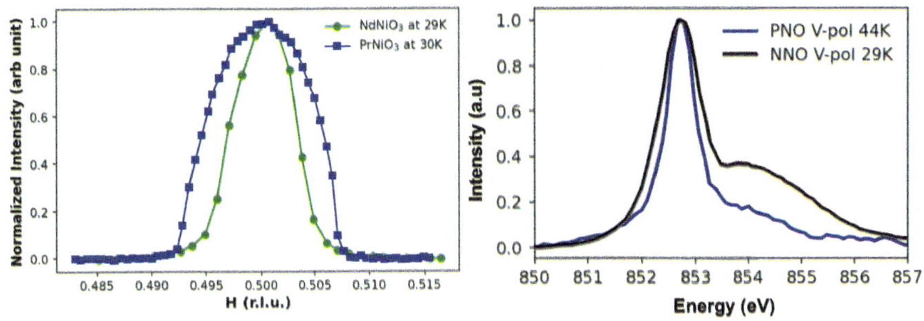

Fig. 3.16 **Magnetic order characterization**: (left) Rocking curve scans on the peak projected onto the H axis is used to extract the correlation length. (right) Fixed-q energy scan on the magnetic peak

Table 3.4 Characterization of the magnetic order

Thin film	Along $q_\perp$ correlation length (nm)	Cooling Néel Temp (K)	$E_B - E_A$ (eV)	Amplitude ratio (B:A)
$PrNiO_3$	29	90	1.4	0.09
$NdNiO_3$	45	210	1.7	0.2

(peak A) and a small Gaussian peak (peak B). In the table below we summarize the characterization of the magnetic peaks for both films. Néel temperature is mentioned for the cooling cycle. The correlation length mentioned is along the $q_\perp$ direction (Table 3.4).

Authors in have thoroughly discussed the response of the magnetic peak energy dependence as a function of the bond distortion. It has been suggested from the calculations that the two peak structure primarily comes from the two different sites where the bond disproportionation drives the effective spin moment on each octahedron. The long bond sites create an effective spin moment close to $S = 1$ whereas the short bond sites host an effective spin moment close to $S = 0$. With an increased bond distortion, the short bond site contributes most to the second peak (peak B) and the long bond to the lower energy peak (peak A). The effect of decreasing bond order is manifested by the short bond peak moving to a lower energy and the long bond peak moving to a higher energy resulting in a decrease in the peak separation. For lower bond order in $PrNiO_3$, we also observe a decrease in the peak separation. The peak amplitude ratio also decreases in the case for $PrNiO_3$ where it is worth noting that we are using vertical polarization to compare the energy dependence. A difference in spin orientation will lead to a decrease in intensity for a given incoming polarization on a given geometry of scattering. This is probably a remnant of what was previously observed.

X-ray absorption spectroscopy: Finally, we explore X-ray absorption spectroscopy collected from the samples in Total Electron Yield (TEY) mode with the

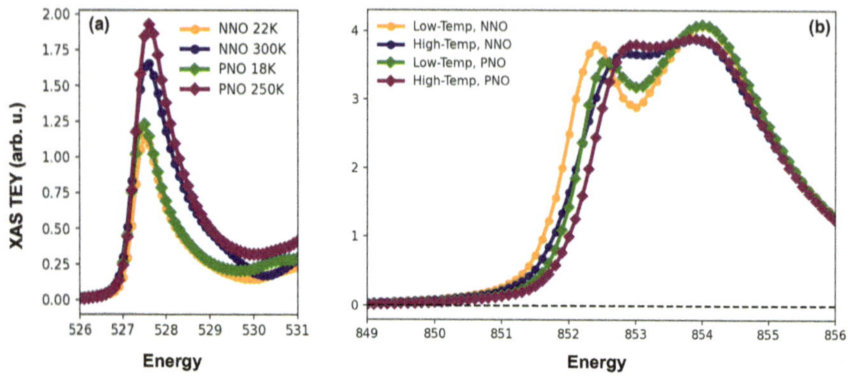

Fig. 3.17 X-ray absorption spectroscopy(XAS) in total electron yield (TEY) mode: (left) O-K edge pre-peak, and (right) Ni-L3 edge for $NdNiO_3$ (NNO) and $PrNiO_3$(PNO) for above and below the transition temperature (T_{MIT})

energy of the incoming beam around Ni L-edge and O K-edge and a resolution of 100 meV. We alternated the polarization between Horizontal and Vertical polarization for each measurement at a given temperature. In the grazing incidence with an angle of 20 deg, the horizontal polarization was mostly perpendicular to the surface of the film while the vertical polarization was entirely in-plane. We measured the ground current and normalized it to a mesh current collected just before the beam entered the chamber. The measured data after background subtraction and averaged over polarization are shown in Fig. 3.17 where we discuss the difference in absorption spectra at those edges between $PrNiO_3$ and $NdNiO_3$ well above and well below the transition temperature.

In the absorption spectra of O-K edge ($1s \rightarrow 2p$), we focus on the pre-peak structure centered around 527.8 eV. The area of the peak is associated with oxygen projected $3d^8\underline{L}$ states and thus an indicator of the Ni-O covalency [1, 19]. With decreasing rare-earth atom size and decreasing tolerance factor (La $\rightarrow$ Lu) the Ni-O-Ni angle decreases from 180°, and so does this peak. Furthermore, it was reported that as $NdNiO_3$ and $PrNiO_3$ undergo the metal-insulator transition the peak area decreases on the high energy side of the peak [19]. The metallic state of $PrNiO_3$ has slightly more Ni-O covalent character than $NdNiO_3$ metallic phase. For both $NdNiO_3$ and $PrNiO_3$, we observe that the pre-peak area under the curve in the insulating state decreases to almost 60% of its high-temperature value. This is interesting since the bond order and charge order in $PrNiO_3$ is much less than that of $NdNiO_3$. This may indicate that the change in the spectral weight during the phase transition in nickelates is independent of the bond order strength.

On the other hand, the Ni edge spectra are consistent with the interpretation that bond order plays a significant role in changing the electronic environment. The primary feature of the spectroscopy data from a nickel L3 edge($2p \rightarrow 3d$) is that it shows a clear two-peak structure. The peaks are less separated at high temperatures

Table 3.5 Ni-L3 edge characteristics in nickelates thin films: $E_{A(B)}$-center of the A(B) peak, $\Delta E = E_B - E_A$, $A : B = E_A/E_B$, $FWHM_{A(B)}$ = full-width at half maximum of peak A(B)

Attributes	$NdNiO_3$ 22 K	$NdNiO_3$ 300 K	$PrNiO_3$ 18 K	$PrNiO_3$ 250 K
E_A	852.3	852.6	852.5	852.7
E_B	854.0	853.9	854.0	854.0
ΔE	1.7	1.3	1.5	1.2
$A : B$	0.65	0.35	0.47	0.37
$FWHM_A$	0.8	1.0	0.8	0.9
$FWHM_B$	2.4	2.8	2.6	2.6

but they appear significantly discernible at low temperature insulating state of the sample. We have tabulated the salient features of the Ni-L3 edge in Table 3.5 after fitting two peaks to the structure. We fit the spectra in the range 851, 855 two Gaussians. Since X-ray absorption spectroscopy is a surface-sensitive technique we are also sensitive to the dirt that the sample has on the surface and which contributes to the background of the measurement.

From the table, we focus on the parameters ΔE, $A : B$, and the width of each peak that are independent of the beamline calibration and background subtraction. To carry out the analysis and subsequent interpretation of data, we follow the work by Disa et al. [20]. They have similarly characterized the data and found qualitative agreement through charge-transfer multiplet modeling. However, it should be noted that the charge transfer multiplet effects cannot cover the peak broadening effects entirely [19]. As the authors noted, the splitting of the Ni-L3 XAS can be directly related to the charge transfer energy between the O-2p and Ni-3d orbitals. A charge transfer energy, as discussed before, relates to the increase of an electron at Ni-e_g orbitals from O-p orbitals and drives the central physics of the nickelates. This creates a ligand hole ($\underline{L}$ in the ligand orbital which remains covalently shared with the Ni-orbital it overlaps with. Such a situation can be interpreted similarly to the Zhang-Rice singlet states in the cuprates, and is responsible for the conductivity.

If the charge transfer energy is increased, it becomes easier for the system to remain covalent. The effect is evidenced by the peak splitting between the two peaks of the Ni-L3 XAS. We name them as A peak(852.7 eV) and B peak (at 854 eV). In the metallic state, the two peaks considerably overlap, indicating a high covalent state of the NiO_6 octahedra. We further note from our extracted value that not only does the peak separation increase but also the A-peak becomes comparatively sharper as the respective nickelates undergo the metal-insulator transition. We can interpret this result from the site-selective Mott transition picture, where two neighboring octahedra assume a short-bond and a long-bond structure. This in turn influences the electronic character from $3d^8\underline{L} + 3d^8\underline{L} \rightarrow 3d^8 + 3d^8\underline{L}^2$. Thus, a decrease in covalency increases the energy separation between the two features corresponding to the two states and makes a more prominent $3d^8$ state observed as the A-peak in our

XAS spectra. The increased separation in $NdNiO_3$ can be attributed to the increased bond order observed in this material, which will enforce a further decrease in the covalent component.

Thus, it is clear from our measurement that the effect of bond order is to decrease the covalency of the NiO_6 octahedra. This is evidenced by the loss of high energy spectra in the O-K edge XAS pre-peak and the increased splitting in Ni-L3 edge XAS. In the L2 edge, a similar low-energy shoulder is observed to be formed due to the metal-insulator transition, in line with the L3 edge. However, since the L2 edge is significantly weaker than the L3 edge we show the L3 edge here for characterization. Furthermore, the change in the electronic configuration may not be discrete and is more continuously tunable as evidenced by the low-temperature XAS ΔE between $NdNiO_3$ and $PrNiO_3$. Furthermore, we note the remarkable similarity between the peak separation of the magnetic peak energy response and the X-ray absorption spectroscopy. We consider that such similarity is a first hint that the effective spin moments of the two Ni sites depend on the bond order [21]. Following Hund's rule, for the $3d^8$ and $3d^8\underline{L}^2$ states the expected moment are $S = 1$ and $S = 0$ respectively. Hence, the ligand holes antiferromagnetically couple with the Ni-holes. As the bond order tunes the covalency of the Ni-O continuously, the effective spin moment also changes continuously, from a $(\frac{1}{2}, \frac{1}{2})$ configuration to (0, 1) in the short-bond and long-bond sites. A gradual tuning from $3d^8$ configuration may result in intermediate spin configurations where the moments are somewhere in between.

Thus, we have explored the charge order, bond order, and magnetic order in rare-earth nickelate materials where the phase transitions are maximally coupled. We find that the magnetic order in $NdNiO_3$ has a longer correlation length compared to $PrNiO_3$. In the energy response of the magnetic peak, we observed the influence of bond order further providing evidence for a lower bond order in $PrNiO_3$.

3.5 Results: Tuning the Phase Transition With He Implantation

In this section, we will discuss the effect of helium implantation on the nickelate thin films. Strong electron-lattice coupling in the nickelates was clearly evidenced in the previous section where a decreased bond order in $PrNiO_3$ compared to $NdNiO_3$ resulted in observing slightly different energy response from the total electron yield data of the X-ray absorption spectroscopy and the fixed-q energy scans of the magnetic order. Since both of these systems show more or less similar characteristics, we study them both under helium implantation. Helium implantation is expected to change the out-of-plane lattice parameter in the thin film, and through this strong lattice-electron coupling we expect to see a change in the electronic configuration and its temperature response. To the best of our knowledge, there has not been a study on nickelates exploring the effect of the uniaxial strain. We will see that, indeed, the helium implantation suppresses the coupled phase transition gradually.

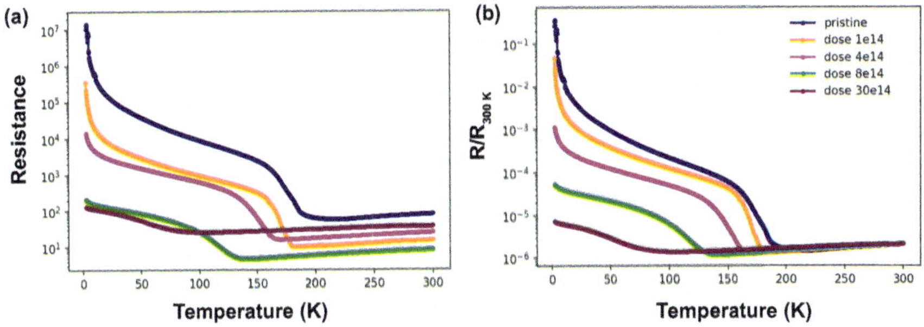

Fig. 3.18 Tuning metal-insulator transition in $NdNiO_3$ films: Four point Van der Pauw measurement on the helium implanted thin films show a gradual suppression of metal-insulator transition. **a** As measured resistance in units of Ω shows the room-temperature resistance drops for the first 4 doses. **b** Normalized resistance with respect to the high temperature (300 K) shows the slope of the high temperature metallic state remains relatively similar

3.5.1 Uniform Helium Irradiation in $NdNiO_3$

To study this, we have identically grown epitaxial thin films of $NdNiO_3$ on 110-oriented $NdGaO_3$ similar to the one discussed above. Then we uniformly irradiated the thin films with helium ions of 90 keV energy. The energy was set to make sure that the ion completely penetrates the thin film and gets implanted deep into the substrate. We exposed different films to different degree of helium ion dose or fluence: 0, 1, 4, 8, and 30 ions per sq. cm. (ions/cm^2) For comparison, we need at least 10^{17} ions per sq cm to create dislocations and damages. Thus, in this limit, the helium will be simply implanted into the substrate.

Tuning of Metal-Insulator Transition

The first important result is shown in Fig. 3.18, where the resistance and resistivity normalized by room-temperature resistivity is plotted for the sample temperature when the sample was warmed from 2 K to 300 K. We measured the transport following the same methodology as the pristine sample discussed above. A gradual lowering of T_{IMT} is observed from the transport data. We note that the slope of the metallic state remains indifferent to increasing helium dose. Also, initially, the raw resistance value at room-temperature decreases with increasing dose up to 8e14 ions/cm^2 indicating a tuning of the conductivity without introducing defects. Probably we introduce some structural defects with a dose of 30e14 ions/cm^2 where the room-temperature resistance is observed to go up.

In Table 3.6 we have listed the transition temperature for the cooling cycle and the warming cycle along with the width of the hysteresis loop for comparison.

We see that the width of the hysteresis loop also remains the same up to 8e14 ions/cm^2. Thus, even with helium implantation, the phase transition remains first order. At the dose of 30e14 ions/cm^2 even though the loop width severely decreases we can still see a clear separation of the warming and cooling trajectories (Fig. 3.19).

Table 3.6 Transition temperatures as extracted from transport measurements

Samples	T_{MIT}	T_{IMT}	Width (ΔT)
pristine NNO 17	215	225	20
dose 1e14 NNO 13	155	180	15
dose 4e14 NNO 12	145	170	15
dose 8e14 NNO 11	120	140	15
dose 30e14 NNO 10	100	100	5

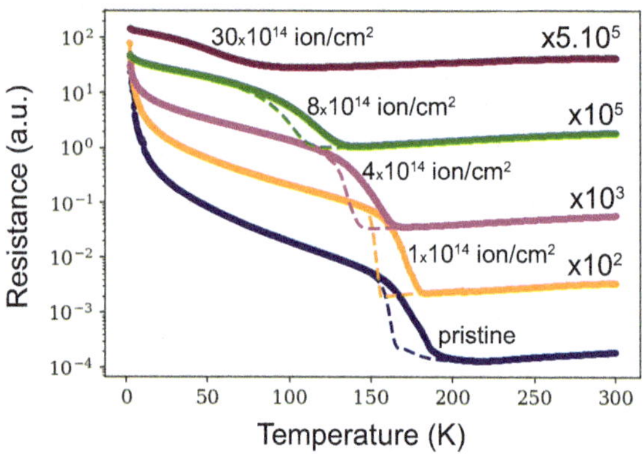

Fig. 3.19 **Influence of helium implantation on the first order character of MIT**: The firm(dashed) lines show the warming(cooling) plots for transport characterization of the films

Structural Order

We investigated the structure of the helium-implanted thin films before and after the helium implantation. Identically grown samples of $NdNiO_3$ hosted identical structures, to begin with. We also used the substrates of the same batch to ensure defect density in the substrate is also roughly similar among the samples. The effect of the helium implantation is similar to the ones reported To characterize the structure we carried out X-ray diffraction on the helium implanted samples using our lab X-ray four circle diffractometer (make Huber) with Cu-K source. The results are shown in Fig. 3.20.

X-ray diffraction: Here we show the reciprocal space map of the peak (221) in HHL plane spanned by the sample normal direction of (HH0) and in-plane direction of (00L). We aligned the substrate peak with the lattice parameters of the pristine sample. The doses increase from left to right. The primary observation is that the thin film peak decreases its HH position without changing the L value as the dose increases. We interpret this as the out-of-plane d-spacing change while the thin film remains epitaxial. Also, note that the main substrate peak doesn't change its position in line with our choice of lattice parameters. Thus, the thin film remains epitaxial in the in-plane directions but changes considerably in the out-of-plane direction. Thus, we are applying a uni-axial strain along (1 1 0) direction.

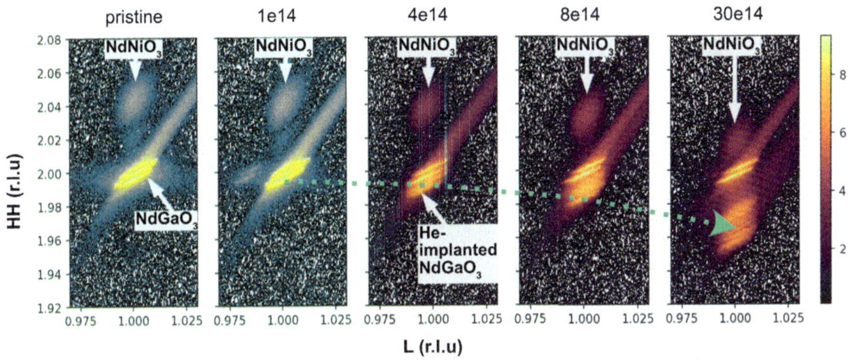

Fig. 3.20 X-ray diffraction on the (2 2 1) peak of the substrate and the thin film: The thin film peak continually decreases in out of plane direction (HH) but remains unchanged in the in-plane direction (L). On the other hand apart from the main substrate peak a distortion peak emerges correlated with increasing helium implantation dose (marked He-implanted $NdGaO_3$ and the evolution traced with a green-dashed line. The doses are marked on top of each panel in units of ions.cm^{-2}

Furthermore, we note that a new 'distortion' peak shows up below the substrate peak. With increasing dose, this peak gains intensity and sinks lower in HH value while also remaining fixed to the L value. *Other references have also reported this.* The helium-implanted layer of the substrate possibly contributes to this intensity while the deep pristine substrate layer remains untouched by the incoming helium from the top, which contributes to the intensity of the original pristine substrate peak position. Since with Cu-K_α, our penetration is much higher we see both of them. To test this hypothesis we do a depth resolve X-ray diffraction on 4e14 ions/cm^2 implanted thin film.

Depth resolved X-ray diffraction: The penetration depth of X-rays in the material depends on the energy, angle of incidence, and composition of the material. For a given material increasing the energy increases the penetration depth. We did a longitudinal scan (th2th scan) on the specular peaks. Both (110) and (220) are symmetry-allowed peaks for the substrate. On the left panel, the data is taken in a soft X-ray chamber of 29-ID beamline of the Advanced Photon Source (APS) whereas the data on the right is collected from a hard X-ray scattering setup at beamline 17-2 of the Stanford Synchrotron Radiation Lightsource (SSRL). Increasing energy increases the penetration depth, in general. We evaluated the exact penetration depth for the incidence angles as 350 nm, 440 nm, 620 nm and 2125 nm from the online CXRO-database for the energies 1700 eV, 2000 eV, 2500 eV, and 8500 eV, respectively (Fig. 3.21).

Here we show that as we increased the penetration depth of the X-rays, the amplitude ratio of the distortion peak to the untouched substrate peak decreased. This confirms the hypothesis we formed from the simulation of ion trajectory. Helium gets implanted near the top of the substrate and modulates the lattice parameters of that substrate layer. To evaluate the helium implanted region near the top surface, we

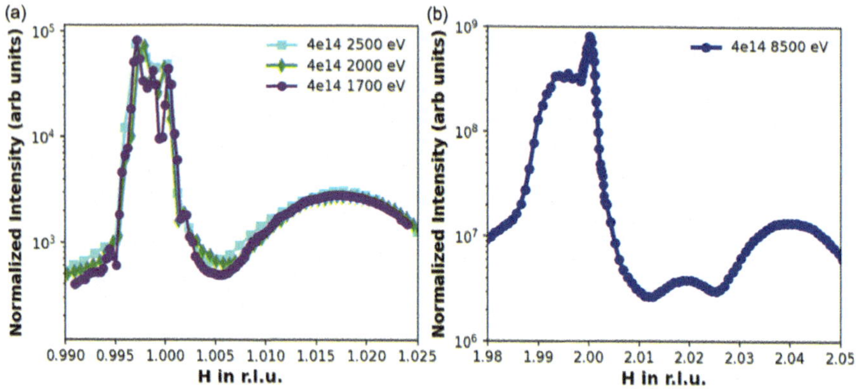

Fig. 3.21 Depth resolved X-ray diffraction on the specular peaks: **a** Data taken in beamline 29-ID of the APS on the (110) peak of the substrate **b** Data taken in the beamline 17-2 of the SSRL on the (220) peak of the substrate. In both scans the film peak is also visible

plot the ratio of the maximum of the distortion peak amplitude to the substrate peak amplitude which shows up at the integer location of the reciprocal lattice unit. The ratio is a direct measurement of the percentage of helium implanted substrate region to the otherwise unaffected substrate regions. As we go deeper, the ratio decreases linearly, indicating an increase in unaffected substrate percentage. Furthermore, the ratio is approximately 1 about the penetration of 1250 nm. Using linear approximation, the top 600 nm of the substrate hosts the implanted helium. Our depth-dependent measurement doesn't exclude the thin film as a host to the helium. However, it is less likely for a high-energy 90 keV helium to be stopped within the overlayer following our discussion in Sect. 3.2.

Our result from the depth-dependent measurement of the depth of helium implantation matches the simulation estimate as well. Since only a thin layer suffers from the helium implantation the in-plane, lattice parameters are kept intact by the rest of the bulk substrate. Moreover, the depth-dependent XRD already shows that the helium implantation effect is inhomogeneous. As the 1700 eV X-ray probes ~350 nm of the top part of the sample, which is well below the range of helium implantation depth, it still records a measurement from the pristine contribution apart from the distortion peak. We will explore the inhomogeneity further (Fig. 3.22).

We have seen so far that the helium gets implanted into the substrate approximately down to 600 nm from the surface, however, the epitaxial relationship remains unperturbed. Thus, under the tensile strain, we expect the distortion in the substrate layer will only change the lattice parameter a and γ. In this case, the d-spacing increases, so γ must decrease. The bending distortion to change γ will propagate along the strain direction into the thin film layer due to the cooperative bending of the connected octahedra. As a result, there will be a subsequent decrease in the γ of the

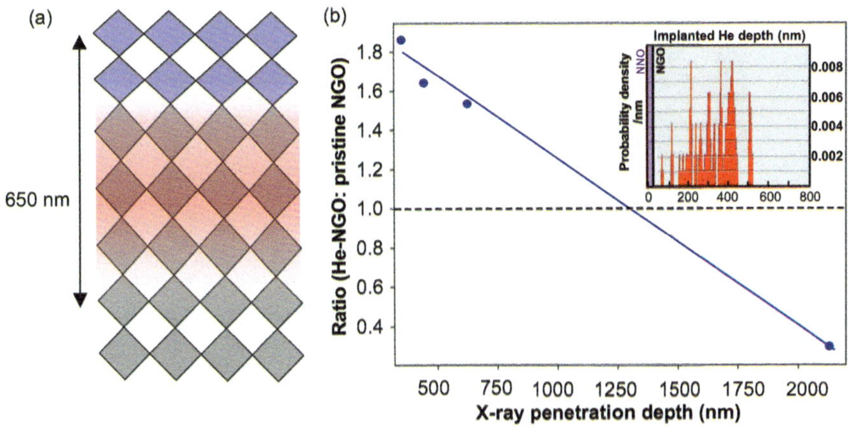

Fig. 3.22 Schematic of structural changes brought about by helium Implantation: **a** A concentrated helium layer is found near the top 650 nm surface of the substrate. **b** From the depth-resolved measurements the ratio of distortion peak to the pristine substrate peak for different probing energy is plotted to find an approximate linear relationship. Where the ratio is 1, half of the scattering volume will correspond to the implanted layer. (*inset* b) the SRIM simulation for 30 ions showing the probability density vs depth confirming implanted ion concentration upto 600 nm

nickelate film resulting in a change in the d-spacing along the out-of-plane direction of (1 1 0). This is observed in the data as the film peak sinks lower in the reciprocal space maps with an increased dose of helium implantation. This is schematically shown in the following Fig. 3.23. We have only shown the lattice plane, where the chemical pressure provided by the helium implantation is mostly adjusted by the co-operative bending and stretching of the octahedra. On the plane spanned by (110) and (001), a stress is managed by changing the c-axis lattice parameter. From the reciprocal space maps it is clear that the c-axis lattice parameter does not change.

To quantify the structural changes experienced by the thin films in this way, we extract the out-of-plane d-spacing on the peak (220), similar to the previous section. Since the helium-implanted substrate layer and the film both remain epitaxial to the untouched substrate layer and under tensile stress, we can use Eq. 3.1. The results are tabulated in Table 3.7 and also in Fig. 3.24.

We note that, to accommodate the uniaxial strain, the γ angle continuously decreases towards 90°, whereas the lattice parameters a and b increase. This is schematically shown in Fig. 3.23. To quantify the amount of uniaxial strain, we assume the definition $strain = \frac{d'-d}{d} * 100$, where d' and d are the d-spacing along 220 for the implanted and bulk $NdNiO_3$, respectively. We summarize the change in lattice parameter and strain along the out-of-plane direction in the Fig. 3.24:

Thus we conclude on the following observations:

- Helium implantation creates a distorted although epitaxially strained layer of the substrate

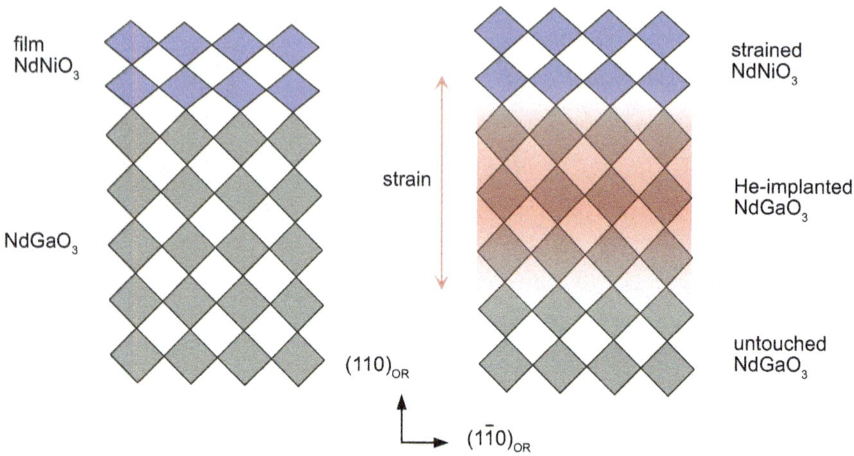

Fig. 3.23 Sub surface helium Implantation generates uniaxial strain: The epitaxial relationship between the substrate and the film helps to propagate the uniaxial strain through cooperative bending of the corner shared octahedra across the interface. Schematically the structural change is demonstrated for (right) helium implanted sample and compared with a pristine film (left)

Table 3.7 Tuning lattice parameters with helium implantation

Attributes	Pristine	1e14	4e14	8e14	30e14
2th (220)	48.16	48.12	48.06	47.86	47.52
a (Å)	5.402	5.404	5.407	5.418	5.436
b (Å)	5.402	5.404	5.407	5.418	5.436
c (Å)	7.73	7.73	7.73	7.73	7.73
α(°)	90	90	90	90	90
β(°)	90	90	90	90	90
γ(°)	91.36	91.31	91.25	91.02	90.64

- Since this layer is in direct contact with our film it changes the strain environment for the film in the out of plane direction only
- To accommodate the change in the out-of-plane direction the lattice parameters of the $NdNiO_3$ film change. However, in this process, the unit cell volume doesn't remain conserved.
- The strain along the out-of-plane direction becomes increasingly compressive ultimately switching sign for the dose of 30e14 ions/cm^2 (Fig. 3.25).

Bond Order and Charge Order

Since we tuned the strain environment and the lattice parameters, an imminent curiosity is how it effects the bond order, and consequently the charge order formation in the nickelates. A hint of decrease is expected from the lowering of the transition temperature and the low-temperature resistance of the insulating phase.

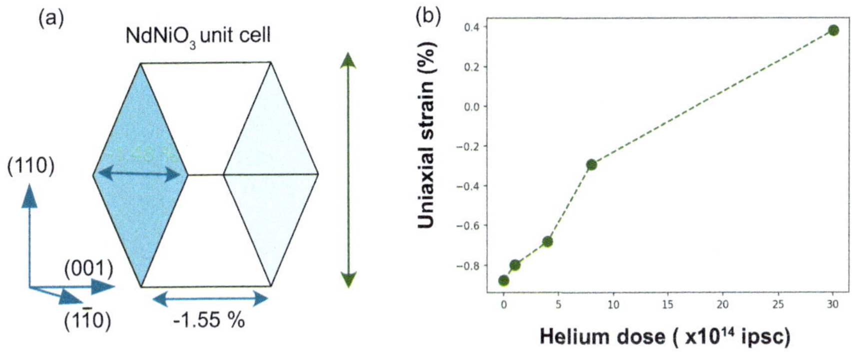

Fig. 3.24 Uniaxial strain tuning by helum implantation: a A schematic of strain experienced by a $NdNiO_3$ unit cell with crystallographic axes noted. **b** The uniaxial strain experienced along the green arrow plotted on the right panel as a function of helium implantation. We observe a clear linear relationship

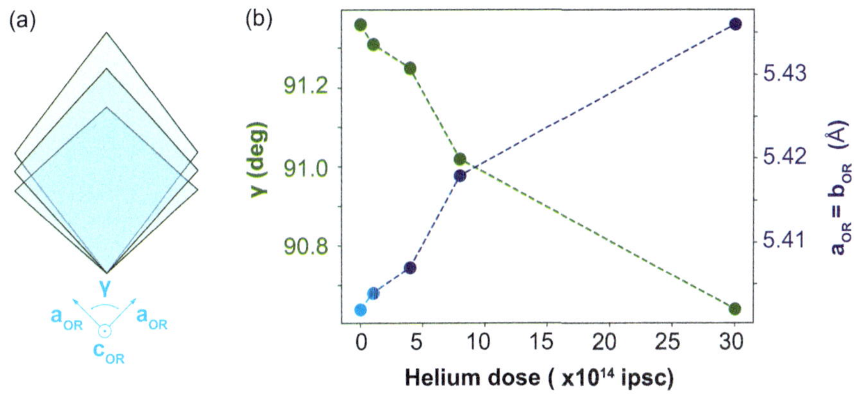

Fig. 3.25 Strain modulates the lattice parameter of $NdNiO_3$ film: a A schematic of the effect of the uniaxial strain on the unit cell shape projected on the c lattice parameter. **b** The lattice parameters evolve continuously with increasing helium dose

To follow up, we investigate bond order and charge order formation in helium implanted $NdniO_3$ using the methodology described in the previous section. We studied the same pristine and helium-implanted samples with doses of 1e14, 4e14, and 8e14 ions per sq cm(ions/cm^2). Due to time constraint, we could not study the sample with helium dose of 30e14 ions/cm^2. However, the data we gathered is sufficient to observe the trend due to the uniaxial strain modulation.

We observe a continuous tuning of the transition temperature of the coupled bond order and charge order transition. The transition temperature decreases continuously. The strength of the bond order also follows a continuous trend with the only outlier of 4e14 ions/cm^2. However, the strength of the charge order does not follow a continuous trend, even though they decrease by 8e14 ions/cm^2 dose. We tabulated the transition

Table 3.8 Bond order transition temperature

Samples	T_{BO}	T_{IMT}
pristine NNO 17	190	225
dose 1e14 NNO 13	175	180
dose 4e14 NNO 12	155	170
dose 8e14 NNO 11	120	140

temperature measured from the warming of the samples along with the insulator-metal transition as observed from the transport data in Table 3.8.

The following Fig. 3.26 summarizes the entire dataset for the warming cycle on each sample under investigation. We note that the data for the sample implanted with 8e14 ions/cm^2 helium shows a sharp artifact around 8.35 keV and possibly due to multiple scattering from the sample. We have collected the cooling cycle data as well which do not differ in general from the warming data, except it is shifted due to hysteresis. For clarity, we have only shown the warming cycle here.

We first note that the peak shape of the energy response does not change much at the low temperature. This could mean that the structure factor relationship between nearest neighbors giving rise to the observed energy modulation remains unchanged, except for their amplitude. The apparent decrease of separation between two peaks in the energy response of the 8e14 ions/cm^2 implanted sample is possibly due to the influence of the multiple scattering peaks and hence regarded as an artifact. Furthermore, we tested the linear relationship of bond order and charge order to test the strength of electron lattice coupling as a function of the strain we provided. We see that the linear relationship is well maintained down to the highest dose measured. Thus, we conclude that the strong electron-lattice coupling in $NdNiO_3$ also remains unperturbed by the uni-axial strain tuning due to helium implantation (Fig. 3.27).

We followed the same energies to extract Δf from the raw data, and subsequently evaluate charge order from it. We have also used the same energy to extract the bond order strength from the raw intensity. We note an overall decrease of bond order with increased uniaxial strain, except for the sample with dose 4e14 ions/cm^2 where both bond order strength and charge order strength increase comparatively. This indicates an incoherent control over bond order by uniaxial strain manipulation. To illustrate this point, we have shown below the temperature evolution of the bond order and the charge order to the uniaxial strain calculated in the previous section (Fig. 3.28).

The decrease in transition temperature, however, indicates that it becomes harder for the system to be in a bond-ordered and charge-ordered phase. Hence, a gradual decrease in the overall strength of bond order fits naturally. If we compare this with the bond-ordered phase of $PrNiO_3$, we note that the slope of charge order to bond order was higher in $PrNiO_3$ and the maximum strengths of charge and bond order were lower. Thus, even though the transition temperature decreases to become like $PrNiO_3$ the electronic configuration must be different from the pristine $PrNiO_3$. This is one of the first evidence we highlight, which indicates that a uniaxial strain tuning

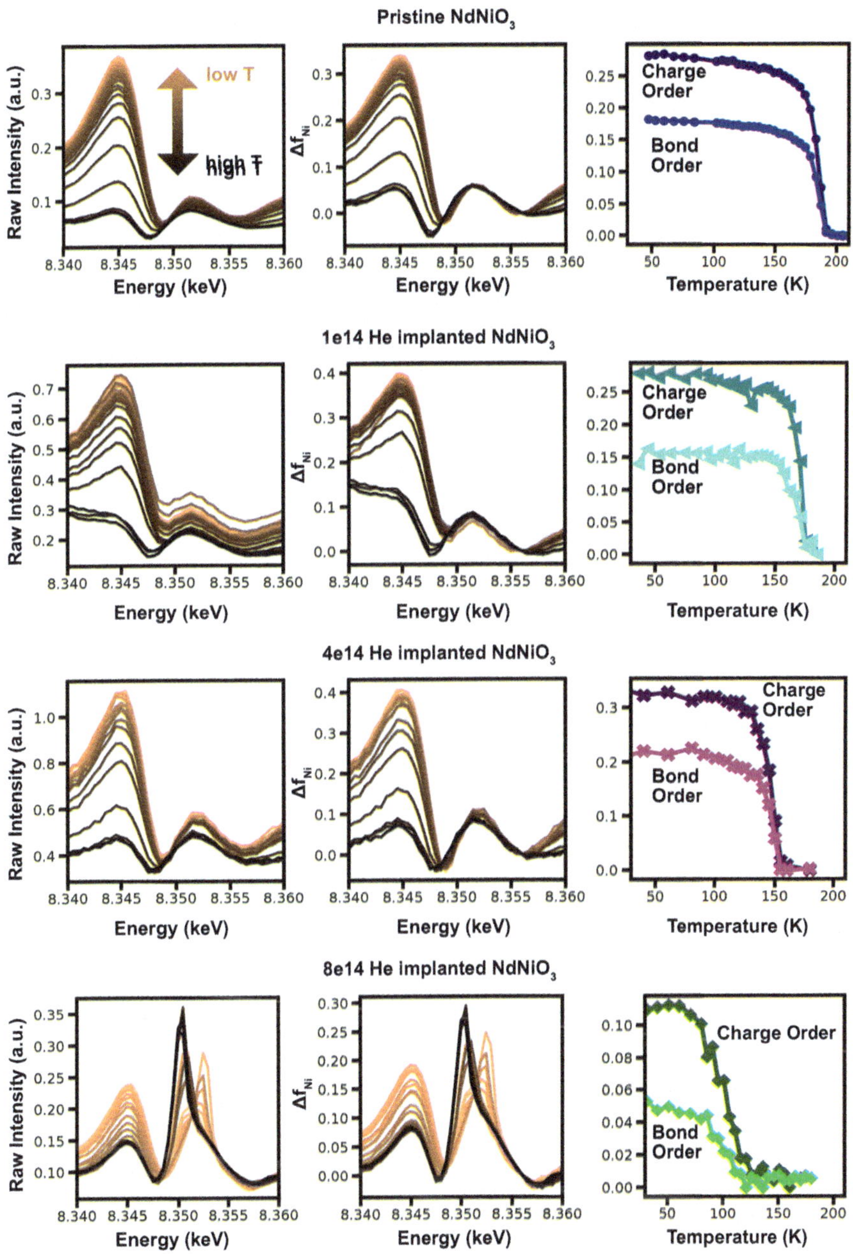

Fig. 3.26 Energy response of the bond order and the charge order and their interrelation: We extracted the bond order and charge order from raw energy response of the $(011)_{OR}$ peak(left most panel) and extracted $\Delta f'$ (middle panel). The temperature evolution of the order parameters are plotted on the right most panel for damages 0, 1, 4, and 8e14 ions/cm^2. Due to time limitation we could not measure the 30e14 sample

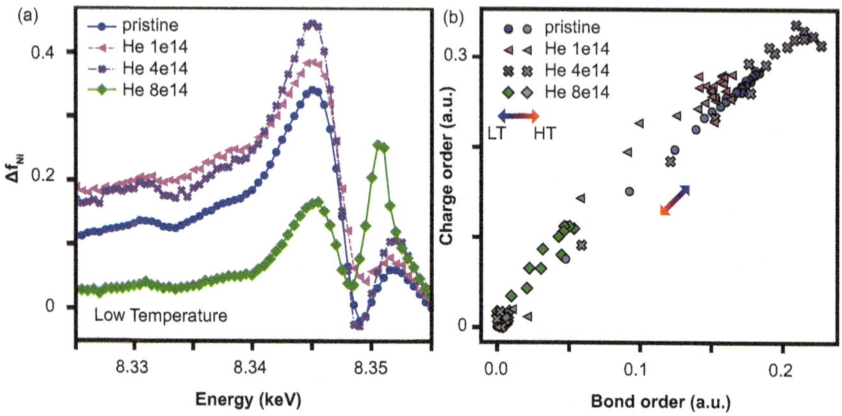

Fig. 3.27 **Characteristics of the bond order and charge order**: **a** Energy response of the resonant contribution $\Delta f'_{Ni}(E)$ **b** Bond order relative to the charge order for evolution for each temperature measured for the warming cycle shows the linear relationship remains valid upto the highest dose investigated indicating that the strong electron lattice coupling is unperturbed. LT = low temperature, HT = high temperature

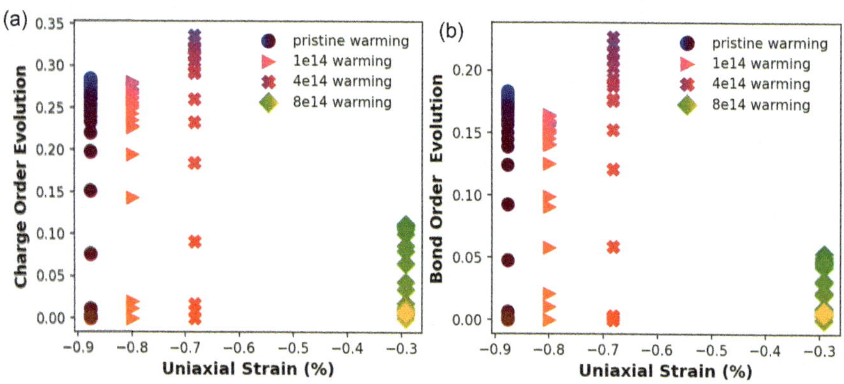

Fig. 3.28 **Evolution of bond order vs charge order with helium implantation**: **a** Charge order **b** Bond order evolution for the warming cycle for the corresponding uniaxial strain. The color grading represents temperature evolution with high temperature collapseing the orders

deviates from the phase space of the nickelate phase diagram in an independent manner. In the following sections, we will find further confirmation of this point.

Finally, we conclude with the following observations from the charge order investigation:

- As expected the bond ordering temperature decreases with an increase in helium dose. This indicates that it becomes harder for the lattice to lower the energy through bond ordering. We observe the strength of the bond order lower with increased uniaxial strain with only the anomaly of 4e14 ions/cm^2.

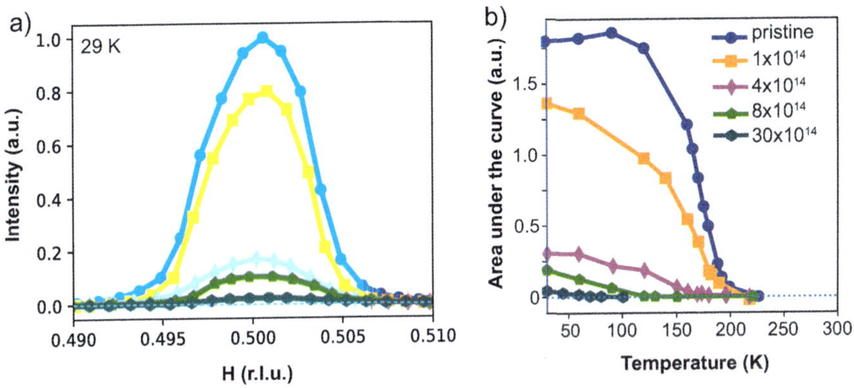

Fig. 3.29 Magnetic peak evolution with increasing helium dose: **a** Rocking curve scan on the magnetic peak shows a decrease in magnetic order strength but relatively unchanged full-width at half max **b** Temperature evolution of the magnetic peak shows a gradual decrease in Néel temperature with increasing helium dose, as expected

- Charge order follows a similar trend as bond order with less prominence. The linear relationship is followed by the charge order as a function of the bond order supporting evidence for strong electron lattice coupling in the nickelate thin film even with the helium implantation in the substrate.

Magnetic Order

Contrasting to the bond order evolution with damage, magnetic order shows a gradual tuning of transition temperature, as well as the magnetic order strength. Such coherent control of magnetism provides significant evidence for spin-lattice coupling in $NdNiO_3$ films. Figure 3.29a shows the lowest temperature magnetic peak as was observed through a rocking curve scan on $q_{AFM} = (\frac{1}{2}, 0, \frac{1}{2})$ in the Kappa-diffractometer chamber of 29-ID of Advanced Photon Source in Argonne National Lab following the same experimental method explained above. With the detector gain remaining fixed, the beam current normalized intensity of the magnetic peak for films exposed with different doses clearly shows a gradual decrease in the magnetic order parameter. We continued to measure the rocking curve scans as we cycled the temperature through the transition temperature on each film. The temperature evolution of the magnetic peak shows a continuous tuning of the Néel temperature as evidenced in Fig. 3.29b. Remarkably, the correlation length of the magnetic phase does not change.

To extract the correlation length of the magnetic order, we fit a Gaussian to extract the correlation length along the transverse direction to the momentum transfer and it was roughly 40–45 nm for all. However, the transition temperature continuously decreases. We show these extracted data in Fig. 3.30. We interpret the result as follows: the domain sizes do not change with increased helium dose but the moment size for each participating atom changes. This interpretation is supported by the

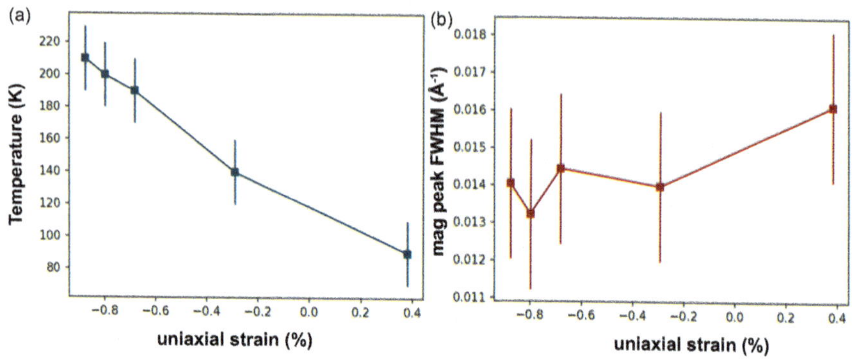

Fig. 3.30 Effect of Helium implantation on $T_{Néel}$ and the correlation length: **a** We see a continuous suppression of the Néel transition temperature **b** Comparatively the peak broadening is not significant

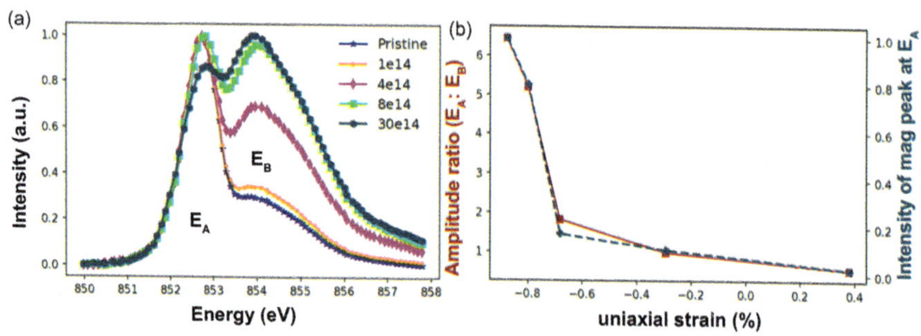

Fig. 3.31 Energy Dependence at low temperature: **a** Energy dependence on the magnetic peak normalized to the maximum of the intensity clearly shows two peak structure we call the first peak (peak A, or) E_A and the second broad one (peak B, or) E_B **b** Intensity from the rocking curve scans plotted together with the ratio of the peaks E_A:E_B directly correlates indicating the drop of the magnetic intensity is correlated with the relative change in the energy distribution

decrease in the scattered intensity on the peak which is proportional to the square of the antiferomagnetic order. The antiferromagnetic order in turn is proportional to the individual moments. On the other hand, the decrease in transition temperature indicates a change in exchange interaction strength between the Ni atoms. To explore more about the change in exchange interaction, we focus on the energy space.

We measured the energy dependence of the magnetic peak at the base temperature of 29 K with fixed-q energy scans. The energy resolution was 0.1 eV. Figure 3.31a shows the energy dependence normalized by the maximum of the detector intensity to highlight the features. We used the right circular polarization to also consider the possibility of spin reorientation due to the change in the electronic environment. Similar to what we observed for the pristine samples the energy response of the magnetic peak is comprised of two peaks: a Lorentzian and a Gaussian. We observe that the peak ratio and the width of the peaks change with increased helium dose.

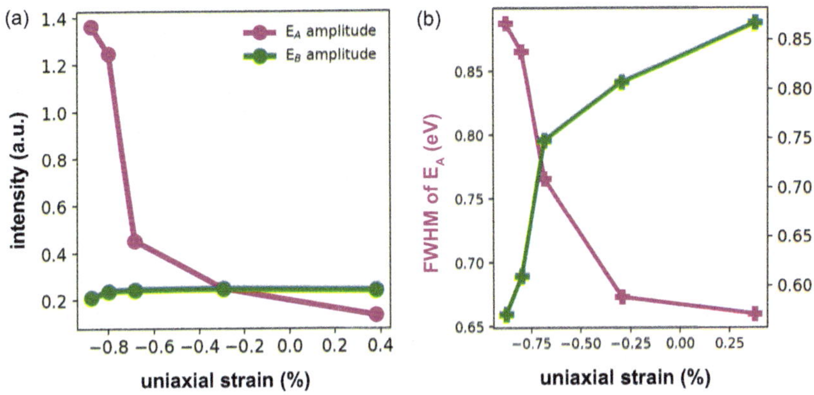

Fig. 3.32 Extracted fit parameters from the energy scans: **a** Amplitude and **b** Full width at half maximum (FWHM) evolution of the energy peaks as a function of uniaxial strain corresponding to the helium dose

Firstly, we measured the intensity of the magnetic peak through rocking curve scans with energy tuned to the first peak (E_A) at 852.7 eV. Figure 3.31b shows the amplitude of the magnetic peak evolution from such rocking curve scans for the uniaxial strain due to the helium implantation. We see the fall of the magnetic peak intensity with decreased uniaxial strain. We simultaneously show the ratio of the two peaks from the fixed-q energy scans. They perfectly match, indicating the fall of the first peak (at 852.7 eV) is correlated with the rise of the second peak at 854 eV.

On closer investigation, we find that the amplitude of the second peak increases slightly while the intensity of the first peak drops sharply. Neither matches the magnetic peak evolution but their ratio. While the peak at E_A becomes smaller and sharper, the peak at E_B becomes broader and stronger. The spread of the energy peaks as extracted from fitting the plots indicates that the broadening of each peak exchanges values as we reach the highest dose measured from the pristine sample (Fig. 3.32).

The continual evolution of the energy response is quite fascinating and demands further investigation. Green et al. thoroughly discussed the energy response from a double cluster model involving two distorted NiO_6 octahedra cluster, based on the site-selective Mott transition picture [21]. The model captures the essential features of the insulating and magnetic phases of the nickelates with negative charge transfer, bond and charge disproportionation, orbital degeneracies, and on-site Coulomb interactions. On top of this, they considered inter-cluster Ni-ligand hopping in their Hamiltonian explicitly. They found that the ligand hole couples with the Ni-hole spin of the e_g orbitals antiferromagnetically and tends to shift to the compressed (short bond) octahedra as the breathing distortion is increased. This continually tunes the effective spin on the neighboring Ni sites, driving the short bond site towards $S = 0$ while the long bond site towards $S = 1$. We have experienced the aspect of continuous tuning of the magnetic moment from our measurement on pristine $NdNiO_3$ and

$PrNiO_3$ magnetic order. However, there is a dynamic charge fluctuation which gets prominent as the bond order is lowered.

As discussed with a strong bond order the ligand holes accumulate at the short bond sites in a tunable manner with the bond order towards the limiting configuration as (4, 2) where the first number in the conjugate pair refers to the total number of holes in the short-bond site and the second number refers to the total number of holes in the long-bond site. Although, even at the strongest bond order, this is not the pure configuration. Due to dynamical charge fluctuations, there is a non-zero configuration weight for the state (3, 3), with a significant configuration weight found to be (4, 2). As we lower the bond order, the fluctuations dominate. At lower bond order, the (4, 2) and (3, 3) can become equal in configuration weight while the (2, 4) also becomes non-zero. At the zero bond ordered phase, (3, 3) state dominate with considerable weight shared by (4, 2) and (2, 4) states. The effect of the charge fluctuation was further found in the energy response of the magnetic peak. The authors showed that even in the limit of zero bond order the charge fluctuations can give rise to a strong magnetic circular dichroism signal. Their calculated energy response shows that increased inter-cluster hopping of the Ni-ligand hole shows up as a split peak structure in the magnetic circular dichroism signal. As the hopping parameter is tuned to be stronger, the peak at the peak position similar to the E_A increases in strength. The resonant magnetic diffraction signal can be found from the magnetic circular dichroism through Kramers-Kronig transformation, which in general preserves the peak positions. Thus, for low bond-ordered samples with a high dose of helium implantation, the two peak structures could be explained by increased charge fluctuations.

They have also found that with non-zero bond order, the long bond site contributes to the E_A peak position and the short bond site to the E_B position. As the bond order strength decreases the two peaks from two different sites come closer until they become indistinguishable at zero strength of the bond order parameter. At this limit, the two peak structure, peak separation, and their relative intensities are found to be maintained by the inter-cluster hopping parameter. Thus, from this discussion, we form the insight that with decreasing bond order inter-cluster hopping and charge fluctuation may take over the electronic configuration in the helium-implanted nickelates and drive the phase transition. However, we do not yet see direct evidence of increased charge dynamics. To see that we will turn towards soft X-ray emission and absorption spectroscopy.

Before turning towards the investigation of inter-cluster hopping, we tie a few loose ends. We note that the peak position of the peak at E_B moves to a lower energy with a decrease in the bond ordering strength. However, the peak position at E_A does not change, as shown in Fig. 3.33. Following the analysis of [21], we note that if the bond order decreases, the two peaks come closer with their energy gap decreasing. On the other hand, the inter-cluster hopping does not considerably influence the E_A peak position. Thus, we can consider a smooth changeover from a strong bond-ordered phase to a high inter-cluster hopping phase. As a result of the shrinking bond order, the E_B peak lowers in energy however, with the increased inter-cluster hopping the E_A peak is sustained. With more strain even the inter-cluster, hopping

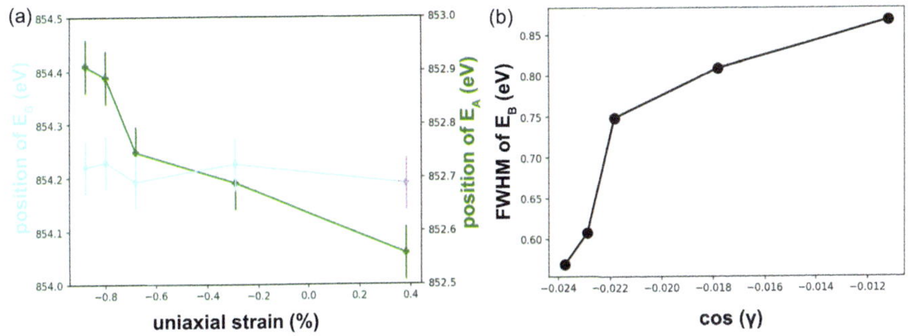

Fig. 3.33 Further characterization of the fits: **a** Peak B position continually decreases and comes closer to peak A position but peak A remains relatively static **b** FWHM of peak B as a function of $cos(\gamma)$

starts to decrease when the Ni-O-Ni distance increases beyond a limit. At this limit, the E_A peak may fall below the E_B peak as seen in double-cluster simulations and the case of damage of 30e14 ions/cm^2.

However, with this evidence, we realize that the electronic environment of the nickelates changes completely with the strain. From a bond order-driven site-selective Mott transition, the magnetic order is driven by charge fluctuations and inter-cluster hopping. In this limit of a highly itinerant hole system, the effect of the Fermi surface becomes important. We plot the increase in the full-width at half-max (FWHM) of the peak at E_B with respect to the $cos(\gamma)$ for each sample and see a correlated relationship. The exact fit for both the FWHM of the peak at E_A and E_B is found and shown in Fig. 3.34. The double cluster simulation does not explicitly discuss the effect on the full-width half max of each peak. Furthermore, the broadening of a peak in a many-body quantum system is influenced by many parameters such as core-hole lifetime, spin and charge fluctuation, coupling to phonon, etc. It is not clear to us if a single parameter or multiple parameters influenced the broadening of the peaks. However, since we consider that the inter-cluster hopping dominates with increasing dose, we can further take the broadening as a signature of the increased hopping. The relationship with $cos(\gamma)$ may not be surprising considering in nickelates Ni-O-Ni bond angle significantly influences the physics.

3.5.2 *Absorption and Emission Spectroscopy*

After we have probed the order parameters of the ground state of nickelates with temperature for each helium-implanted sample, we observed that the helium-implantation generated uniaxial strain tuned the coupled paramagnetic-metal to bond and charge-ordered antiferromagnetic-insulator transition. From the energy dependence of the magnetic peak, we realized that the electronic character of the transition

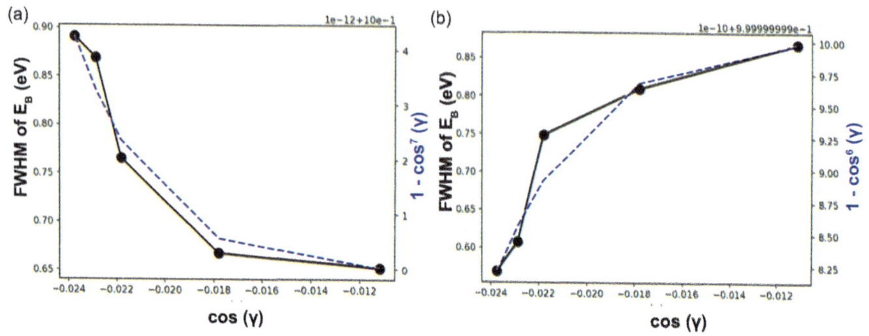

Fig. 3.34 **Width of energy dependence and its link with** $cos(\gamma)$: Full width at half maximum of the energy response at **a** E_A, and **b** E_B location and how they vary with cos of unit cell angle, γ due to uniaxial strain

is also changing from a site-selective Mott transition (where strong bond order drives the transition) to a transition where bond order is low and dynamic charge fluctuation is high. To investigate further about the electronic configuration we investigate here room-temperature and low temperature X-ray absorption spectroscopy (XAS) in total electron yield (TEY) mode. The experiments were carried out in Beamline 4.0.3 of the Advanced Light Source (ALS) and followed the methodology described in the previous section for the pristine samples.

The raw data were similarly background subtracted and normalized to produce the plot in Fig. 3.35, where we have plotted the room-temperature XAS TEY for O-K edge pre-peak and the Ni-L3 edge. From the O-K edge we see that the pre-peak shrinks with increased helium implantation. This pre-peak is considered the oxygen orbital projected $3d^8\underline{L}$ states and an indicator of the Ni-O covalency. Thus, we have direct evidence that with helium implantation we are changing the electronic configuration of the nickelate even at the room-temperature metallic state. The change in the Ni-L3 edge is more subtle. We see a typical metallic XAS spectra with two peaks almost merged together (E'_A, and E'_B) This slight change will come in handy while studying pheotoemission electron microscopy on focused ion samples. We have multiplied the polarization-averaged background-subtracted spectra with a series of numbers to show that the intensity of peak A (E'_A) increases slightly.

We have further extracted the quantities that describe the peaks in the above XAS and have shown in the Fig. 3.36. Firstly, the area under the O-K pre-peak within energy 526–530 eV is discussed first. We can see that the area decreases to ~60% of the pristine state. For pristine $NdNiO_3$ and $PrNiO_3$, we observed such shrink in the area under the O-K edge pre-peak for metal-insulator transition. However, that was accompanied by the peak splitting of the Ni-L3 edge to 1.7 eV gap between the two peaks. We see that ΔE from the pristine 1.3 eV value decreases by 0.06 eV only. We note here that the resolution for the measurement was ~0.1 eV and thus the shift falls below the resolution of the measurement. Thus, even the room-temperature state of

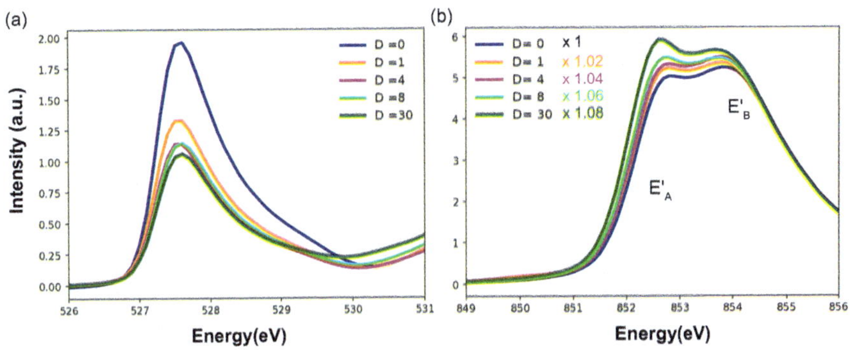

Fig. 3.35 **Room temperature XAS**: X-ray absorption spectroscopy in Total electron yield mode in grazing incidence with polarization averaged for helium implanted samples at room-temperature **a** O-K edge pre-peak **b** Ni-L3 edge, after background subtraction and normalization

the helium-implanted sample is unique. Thus, the tuning with helium implantation has opened a new dimension of phase space for the nickelates.

Secondly, from the fit of two Gaussian to the Ni-L3 XAS intensity we note that the full-width at half-max(FWHM) for peak A roughly remains constant or decreases ever so slightly, whereas the peak B FWHM increases indicating a broadening. For any helium-implanted sample including the pristine, there is no bond order at room-temperature and we observed a decrease in the $3d^8\underline{L}$ state thus the change in the peak ratio must be brought in by the increased charge transfer energy accompanied by the decrease in covalency. However, the increase is comparatively lower than what we observed in the insulating state again providing support for a distinct state compared to the pristine metallic or insulating state of the sample. In this state describing the physics locally is not enough.

Finally, we extracted X-ray linear Dichroism (XLD) and from that, we extracted the hole ratio between the out-of-plane and in-plane d orbital. We note that there is a slight change in the hole distribution but eventually with a higher helium dose the distribution becomes isotropic.

Next, we turn our attention to the electronic configuration of the unoccupied states as we lower the temperature. All of the samples undergo a metal-insulator transition, however, at a different temperature. Increased doses result in lower transition temperatures. We have collected the data at 22 K where all samples are insulating according to the transport measurement.

We see that the O-K edge pre-peak shrinks even further. The extracted area under the curve is shown in Fig. 3.37 along with the room-temperature data. The area under the curve for any sample at any temperature is normalized by the low-temperature pristine data. It is interesting to observe that as the implanted sample undergoes the transition they decrease the covalency further to an extent that is similar to the pristine phase transition. Without the bond order, it seems the system tries to find the ground state by decreasing the covalency. As if it knows only one way to decrease the

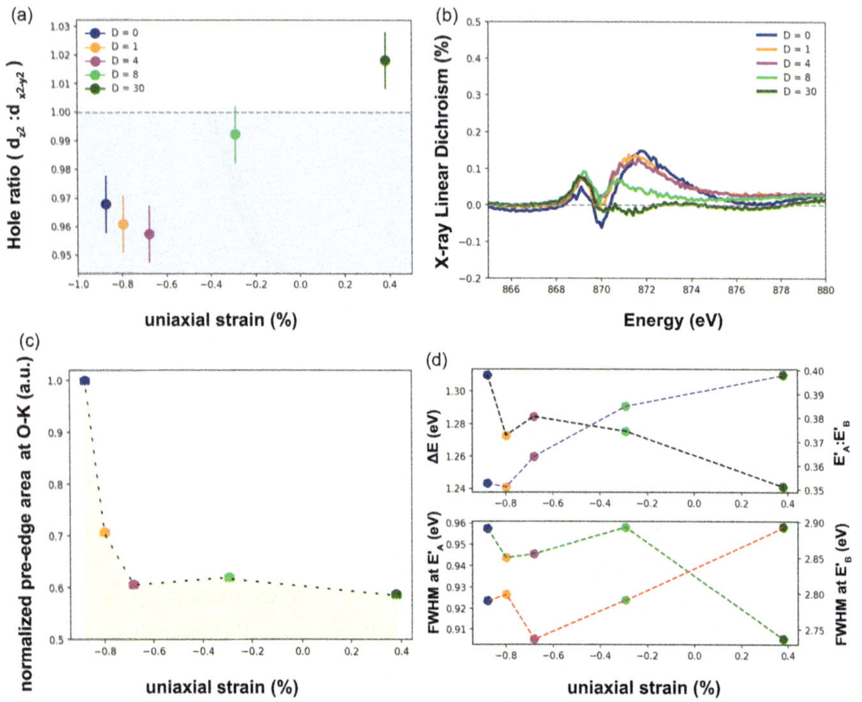

Fig. 3.36 Room temperature Characterization of the XAS: Peak characteristics extracted from the room-temperature XAS **a** Anisotropy of hole ratio **b** X-ray Linear Dichroism on the L_2 peak **c** Area under the O-K pre-edge peak **d** Detailed characterization of the two peaks (E'_A and E_B) at L_3

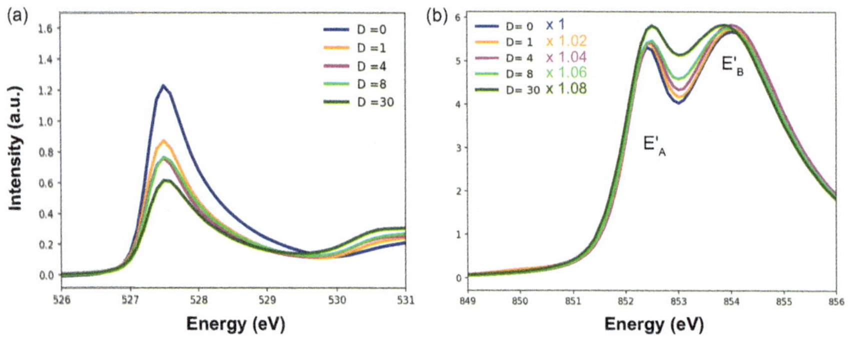

Fig. 3.37 X-ray absorption spectroscopy in Total electron yield mode in grazing incidence with polarization averaged for helium implanted samples at low temperature (22 K) **a** O-K edge pre-peak **b** Ni-L3 edge, after background subtraction and normalization

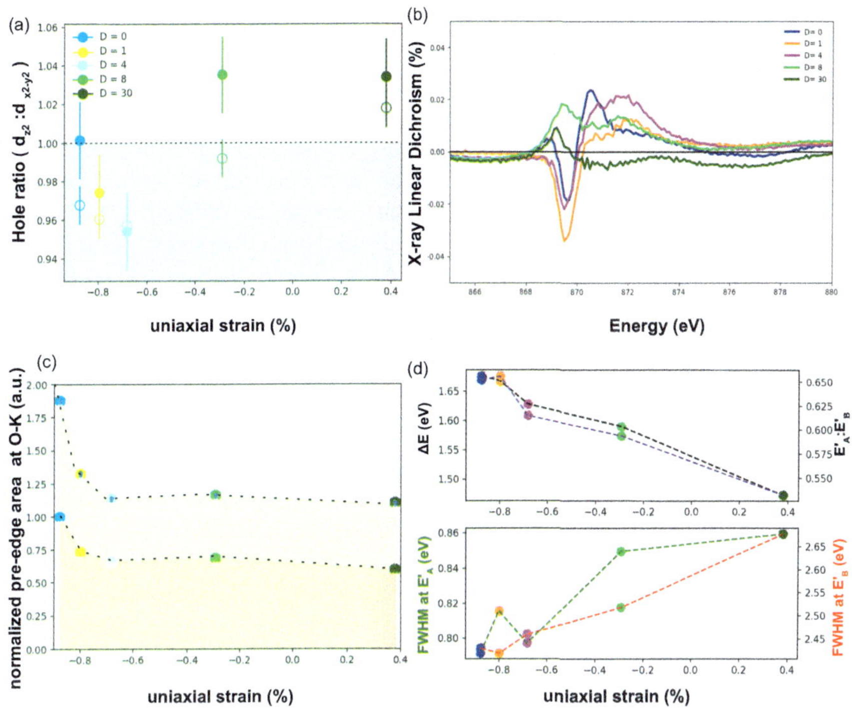

Fig. 3.38 Low temperature Characterization of the XAS and comparison with room temperature data: Peak characteristics extracted from the low temperature XAS **a** Anisotropy of hole ratio **b** X-ray Linear Dichroism on the L_2 peak **c** Area under the O-K pre-edge peak **d** Detailed characterization of the two peaks (E'$_A$ and E_B) at L_3

overall energy, however, one question remains: What means does it take to decrease the covalency as bond order is unavailable?

On the other hand, we observe familiar two-peak splitting for the low-temperature Ni-L3 edge peak; as expected from decreased covalency and increased charge transfer energy. With increased helium dose and hence uniaxial strain the peak splitting decreases slightly. Both peak A and peak B broaden and the peak ratio of $A : B$ decreases. If we compare the high temperature ΔE with low temperature ΔE, we realize that the change is the largest for the pristine sample and decreases continuously up to the highest helium dose we explored. Thus, even though the O-K edge pre-peak shrinks similarly, the peak separation at the Ni-L3 edge does not undergo similar change due to the metal-insulator transition. Furthermore, we observe the sharpness of the separated peak decrease with increased uniaxial strain from the implanted helium. There is intensity between the two peaks that increases in weight as helium implantation increases. The existence of these new states is echoed in the increased broadening of the peaks and decreased peak ratio. In the table below we summarize the detailed finding of the XAS data for both temperatures (Fig. 3.38).

X-ray absorption spectroscopy told us that both the room-temperature and the low-temperature states differ significantly from the pristine metal and insulator state, in terms of electronic configuration. Furthermore, the calculations of local approximations seem to fail to explain the evolution of the XAS with helium implantation such as the pre-peak of the O-K edge decrease but the Ni-L3 edge does not split simultaneously. Thus, it is imperative to investigate the occupied electronic states near the Fermi level for Ni valence shells. We thus conduct an energy-resolved X-ray emission experiment at the resonance edge of the Ni-L3 also known as Resonant Inelastic X-ray Spextroscopy (RIXS).

To conduct the experiment, we mounted all of the samples simultaneously on the sample holder. The incident angle of the incoming beam was set to 30° and the scattering angle was fixed at 150°. Emitted photons passed through a grating before being recorded by a CCD detector (make Andor) of 2048 X 128 pixels. The energy of the emitted is resolved along the long direction and the short direction is summed over for statistics. Before the start of the experiment, we tested the instrument resolution using elastic scattering from carbon tape in the energy range we intend to collect the data. The overall energy resolution was found to be ~700 meV at 50 μm exit slit size. The energy resolution is slit size-dependent. We obtained a lower resolution at ~500 meV, however, the intensity dropped quite a bit and for the sake of limited time, we decided to continue with the 700 meV resolution.

At this scattering geometry described above we vary the energy of the incident beam from 852 eV to 861 eV with a step size of 250 meV. From the collected spectra we subtracted the incident energy to plot the energy transfer to the sample versus incident energy, called the energy map. This data is shown for all of the helium-implanted $NdNiO_3$ samples. We begin the discussion of the energy map following the pioneering work of Bisogni et al. [7].

In the low-temperature insulating state, two distinct excitations were found in similarly epitaxial $NdNiO_3$ thin film: localized *dd* excitation around 1 eV energy transfer, and fluorescence excitation (FL) around 2–3 eV which was reported to have an itinerant component [7]. The itinerancy of an excitation can be read from the energy map if with increased excitation the energy transfer linearly increases. For the pristine sample, we reproduce the result of Bisogni et al. only with a lower resolution. Despite the resolution, the primary features can be immediately seen. The localized *dd* excitation is found around 1 eV quite separated from the fluorescence excitation around 2 eV which has an itinerant component. The incredible observation comes from the subsequent energy maps where, increasing the helium-implantation localized component of the *dd* excitation starts to merge with the fluorescence excitations. The gradual tuning of the delocalization of the *dd* excitation continues up to the highest explored implantation dose where both excitations merge along the itinerant line (Fig. 3.39e, f) and no clear distinction exists between the florescence excitation (FL) and the *dd* excitation any more. This precisely provides evidence that the low-temperature insulating state is very distinct in the electronic configuration of the implanted samples. They continue to become different with increasing doses. Furthermore, this explains the density of states in between the A-peak and B-peak that were observed in the XAS TEY measurement (Fig. 3.40).

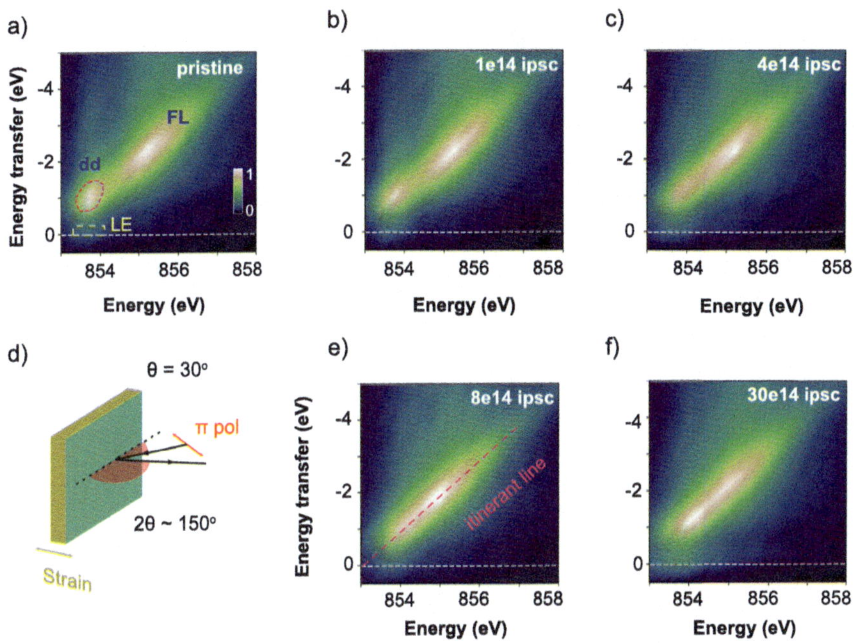

Fig. 3.39 Resonant inelastic X-ray scattering (RIXS) measured at 78 K: The dd excitation becomes increasingly delocalized for higher helium ion implantation and merges with the charge transfer excitation. This also increases the low energy excitation (LE) between 0 and −1 eV indicating a poorer insulating phase for the helium implanted samples. Other important part of excitation is shown in panel (**a**). Experimental geometry is shown in panel (**d**). **e** shows two of the most important character of resonant excitation—localized, with fixed energy transfer value for defined incoming energy, and itinerant, with linearly increasing energy transfer for higher incoming energy

The Fig. 3.39 provides further direct evidence about the increased itinerant character of the nickelate physics. The calculation by [21] showed that the bond order competes with the charge fluctuation, thus, with an increased bond order, we did not see the itinerant component of the nickelate physics. One limit of the physics was suggested by [22], where an itinerant component of the Hamiltonian results in a Fermi surface where a spin density wave instability exists. As we have seen previously, such an evaluation correctly extracts the wave vector of the antiferromagnetic order. Furthermore, in systems with more itinerant characters, such an explanation seems to prevail. For example, in $PrNiO_3$-$PrAlO_3$ superlattice, a weakly metallic state with strong antiferromagnetic order but weak charge order was reported, which could be readily explained by the density wave theory which considers antiferromagnetic order as the primary order parameter and charge order as a result of the magnetic order. For such an explanation, the existence of bond order is not necessary. With the increased itinerant character of the d-electrons of Ni, we might as well be in this configuration space where the phase transition can be explained by the emergence of a spin density wave order.

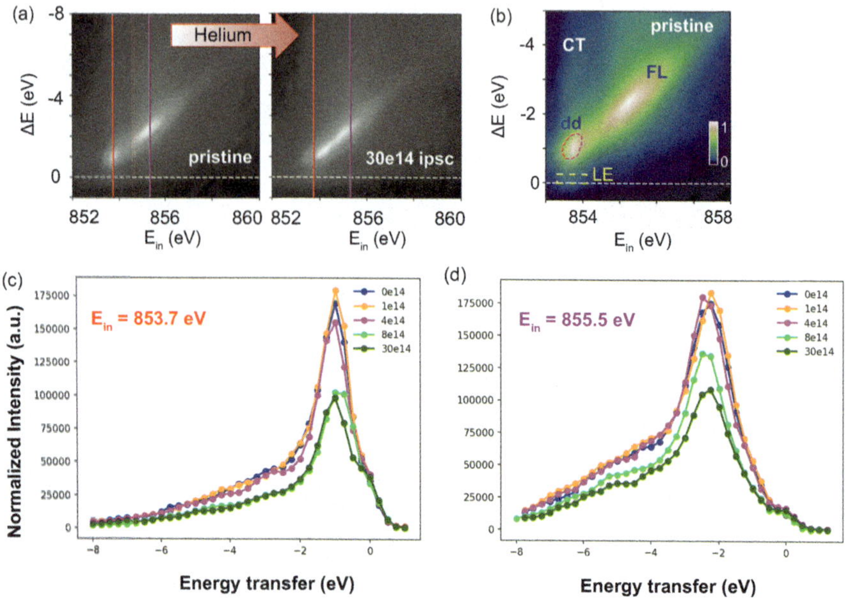

Fig. 3.40 **Decrease in dd excitation energy and FL and CT energy**: **a** Color coded cut locations for the plots in panel (**c**, **d**), **b** Shows the definition of relevant excitations **c**, **d** Shows decrease in the charge transfer excitation with increased helium implantation for equal strength of elastic intensity

Finally, we end this discussion by pointing out that the itinerant character of the d-electron increases, while the low energy excitations below the 1 eV energy line get more weight with increased implantation. It becomes visibly clear in the energy map of 30e14 ions/cm^2. Thus not only do the high-energy excitations change the characteristics, but previously absent low-energy excitations show up as a result of uniaxial tuning. We have taken slices of the energy map at the incoming energy aligned to the *dd* peak location and to the center of the charge transfer excitation. For the doses of 8e14 and 30e14 ions/cm^2 the increase in low energy excitation below the *dd* excitation and above the elastic line can be seen. The broadening of the high energy excitation to the low energy side makes the system more metallic. Thus, with considerable broadening of the high energy excitation we can expect a metallic system with antiferromagnetic order. Since the bond order was stronger in $NdNiO_3$ to reach the limit with low helium dose such that we remain inside the region where defects are experimentally not observable, we use $PrNiO_3$ as the tunable material.

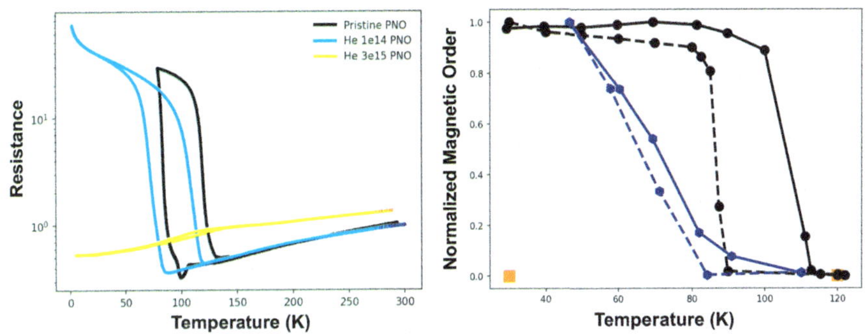

Fig. 3.41 Helium implantation in $PrNiO_3$: Metal-insulator transition and paramagnetic to antiferromagnetic transition both are simultaneously suppressed

3.5.3 *Uniform Helium Irradiation in $PrNiO_3$*

Now that we have realized the effect of helium ion implantation effect on $NdNiO_3$ we look for the applicability in the case of $PrNiO_3$, where the bond order in the pristine state was weaker than $NdNiO_3$. In $PrNiO_3$, the lattice electron coupling was strong, so the strain tuning is expected to similarly tune the electronic state of the sample. We were successful in suppressing the phase transition however the decrease of transition temperature was not smooth and collapsed rapidly. Ultimately, in the process of tuning the phase transition, we discovered metallic antiferomagnetism in $PrNiO_3$ thin films.

Tuning Metal-Insulator Transition

We highlight the tuning of the metal-insulator transition in 22 nm $PrNiO_3$ thin films epitaxially grown on 110 oriented $NdGaO_3$ substrates through Fig. 3.41 where we have shown that the coupled paramagentic-metal to antiferromagnetic-insulator is suppressed. However, it took much less helium dose to suppress the transition completely. By the dose of 30e14 ions/cm^2 (the highest dose for the case of $NdNiO_3$) the sample was completely metallic with no antiferromagnetic order at $\mathbf{q} = (1/2, 0, 1/2)$. For a dose of 4e14 ions/cm^2 we observed no antiferromagnetic peak as well. Thus, we realize for $PrNiO_3$ the control by helium implantation becomes more difficult (Fig. 3.42).

We considered in the low bond ordered pristine state the overlayer lattice might not be rigid enough to sustain tunability by the uniaxial strain helium implantation provided. We then tried a new thing. We decreased the thickness of the overlayer to make the epitaxial strain from the substrate more homogeneous on the $PrNiO_3$ film. A thinner film has a stronger epitaxial relation with a decreased strain gradient compared to a thicker one. The trick helped us to tune the metal-insulator transition in $PrNiO_3$ thin films of 15 nm thickness. We implanted those 15 nm thick $PrNiO_3$ films with helium with doses of 0, 1, 2, 3, 4 and 5e14 ions/cm^2. We show the result of continuous tuning of the metal-insulator transition in these films in Fig. 3.43.

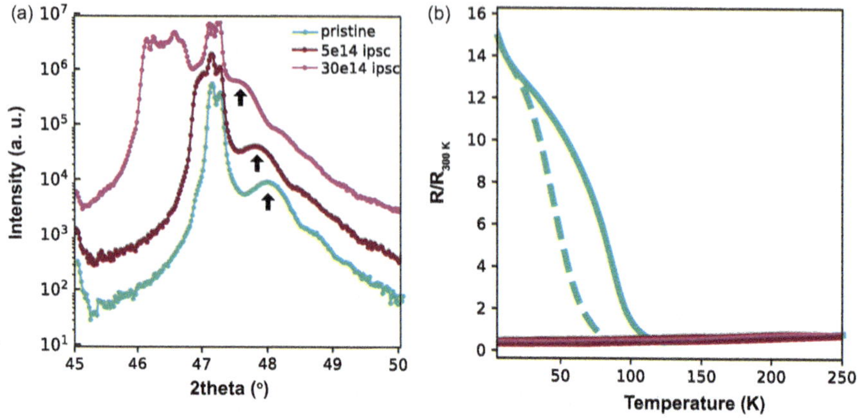

Fig. 3.42 **Gradual Tuning of the of the strain and MIT in 15 nm thick $PrNiO_3$:** **a** $\theta - 2\theta$ scnas on (220) peak showing expected expansion of the film peak marked by black arrow **b** Suppression of the metal-insulator transition even with 5×10^{14} ions/cm^2 dose

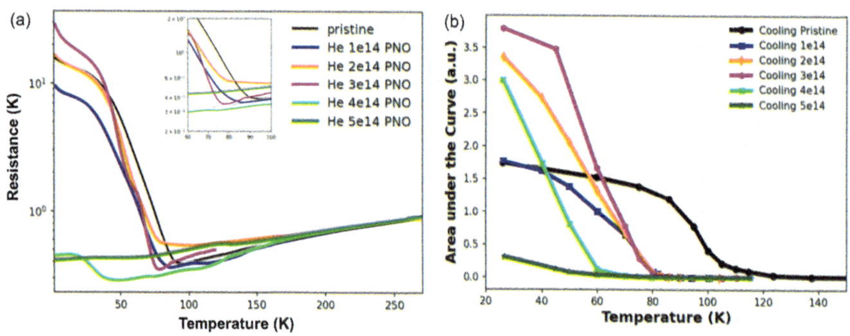

Fig. 3.43 **Suppression of MIT in 15 nm thick $PrNiO_3$:** temperature dependent **a** Transport characterization **b** Antiferromagnetic order from resonant soft X-ray scattering at Ni-L_3 edge shows irregular control of the MIT for doses 1, 2, 3 $\times10^{14}$ ions/cm^2. However for 4 and 5 $\times10^{14}$ ions/cm^2 they again become gradually tunable

The four-point van der Pauw transport measurement showed a continuous tuning of the phase transition. Although we could continuously tune the transition temperature, the control was not as nice as $NdNiO_3$. We see that room-temperature normalized resistance suddenly jumps for the dose of 3e14 ions/cm^2 only to sufficiently collapse for the next higher dose of 4e14 ions/cm^2. Remarkably, the resistance of the sample with 5e14 ions/cm^2 dose does not change the slope of the resistance versus temperature curve down to the lowest temperature, deeming it as a metal. However, the resonant soft X-ray scattering measurement finds an antiferromagnetic peak in this sample. The control over antiferromagnetic transition showed an incoherent control similar to the transport measurement. Thus, the sample implanted with 5e14 ions/cm^2 shows undeniable evidence of being an antiferromagnetic metal.

Antiferromagnetic Metal

Let us focus our attention on the sample implanted with 5e14 ions/cm^2 which appears to be a metal from the transport measurement and an antiferromagnet (AFM) with $\mathbf{q}_{AFM} = (1/2, 0, 1/2)$ from the resonant soft X-ray elastic scattering. We further characterized the AFM peak with rocking curve scans. The extracted full-width at half-max(FWHM) is plotted along with the rocking curves in Fig. 3.44. We have also shown the continuous evolution of the Néel temperature with increased helium dose. Similar to the $NdNiO_3$ sample, we observe the correlation length ($\sim\frac{1}{FWHM}$ remains unchanged with increased doses of helium implantation. All of the changes in the magnetic order are thus accounted for by the lowering of the magnetic moment at each Ni site. The transport characterization, for complex material like nickelates, can be misleading. For instance, if the sample has inhomogeneous domain distribution at low temperatures without filling the volume of the thin film, then scattering from the sample will show an antiferomagnetic peak. However, there can be conducting pathways avoiding the insulating domains. It is worth noting that the magnetic domain sizes do not differ from the pristine sample indicating such domain arrangement leading up to a metallic behavior is not applicable for the $PrNiO_3$ sample. We still look for a local electronic signature that will prove beyond doubt the metallicity of the sample. We have collected fixed-q energy scan data on the magnetic peak of the sample. Moreover, while collecting this data we have collected the Total Electron Yield through the Ammeter connected to the sample. This data is shown in Fig. 3.45. In the energy scan of the magnetic peak, we observe an initial sharpening of peak A with a gradual lowering of peak B intensity. However, after the dose of 3e14 ions/cm^2, the trend reverses, finally reaching the metallic sample of 5e14 ions/cm^2 helium dose where both peak are of equal intensity. Corresponding TEY measurement shows the signature two peak structure of the insulating nickelate state for samples upto 3e14 ions/cm^2 dose. It decreases for 4e14 ions/cm^2 and for 5e15 ions/cm^2 implanted

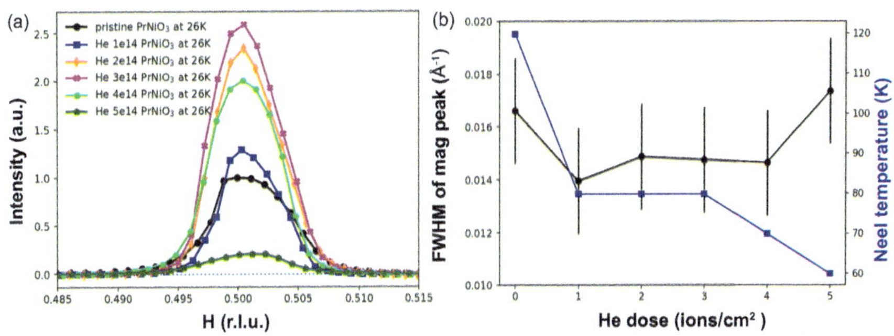

Fig. 3.44 Evolution of the antiferromagnetic order with increased helium implantation in 15 nm $PrNiO_3$: a Rocking curve scans on the AFM peak shows effect on amplitude of the Bragg peak due to helium implantation. The full width half max (FWHM) appears unperturbed. **b** Extracted FWHM from gaussian fits to the plots and the T $_{Néel}$ as a function of helium dose

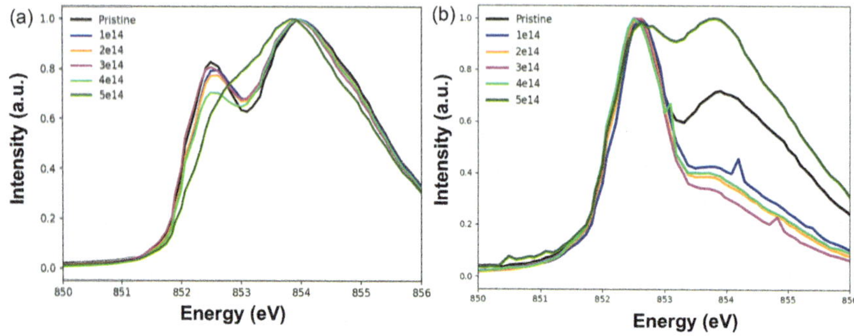

Fig. 3.45 Energy response measured on the antiferromagnetic peak: (left) Total electron yield in grazing incidence polarization averaged (right) Fixed-q energy scan on the magnetic peak

sample we do not see the two peak structure of the XAS. Instead, we see a response usually observed in the metallic state of $PrNiO_3$.

Thus, similar to compressively strained $PrNiO_3$-$PrAlO_3$ superlattice we consider the sample implanted with 5e14 shows a metallic spin density wave state.

3.5.4 Key Understanding from Uniform He-Implantation

We observe a tunable itinerant character of the dd-excitation in the low-temperature state. A likely origin of this delocalization is enhanced inter-octahedral charge fluctuations. In the pristine metallic state, the Ni orbitals are delocalized by virtue of strong hybridization. In the site-selective Mott transition picture, the insulating state hosts two neighboring octahedra with primarily $3d^8$ and $3d^8\underline{L}^2$ characters. The latter forms a singlet that is localized and detached from the neighboring octahedral network. On the other hand, the standalone $3d^8$ state can extend only as much as the orbital wave functions allow, making it effectively localized as well (Fig. 3.46a). In pristine nickelates, these two configurations result in an insulating ground state. However, applying uniaxial strain significantly alters the spectroscopic character of the ground state and reduces the bond order parameter. Thus, the strain enhances the d-orbital itinerancy, evident in the emergence of itinerant dd excitations, while simultaneously reducing Ni-O covalency. The presence of these excitations indicates delocalized d-electrons even in the ground state, illustrated schematically by diffused blue clouds representing an extreme itinerant case.

We thus identify uniaxial strain as a tuning axis orthogonal to the conventional nickelate phase diagram, promoting a more itinerant, less covalent ground state. This evolution is captured in Fig. 3.46b as an extended phase diagram incorporating uniaxial strain as an independent control parameter. We show that uniaxial strain in as-grown nickelate films drives a weakly interacting itinerant regime with a modified antiferromagnetic state that is distinct from the Mott-like phases and complementing

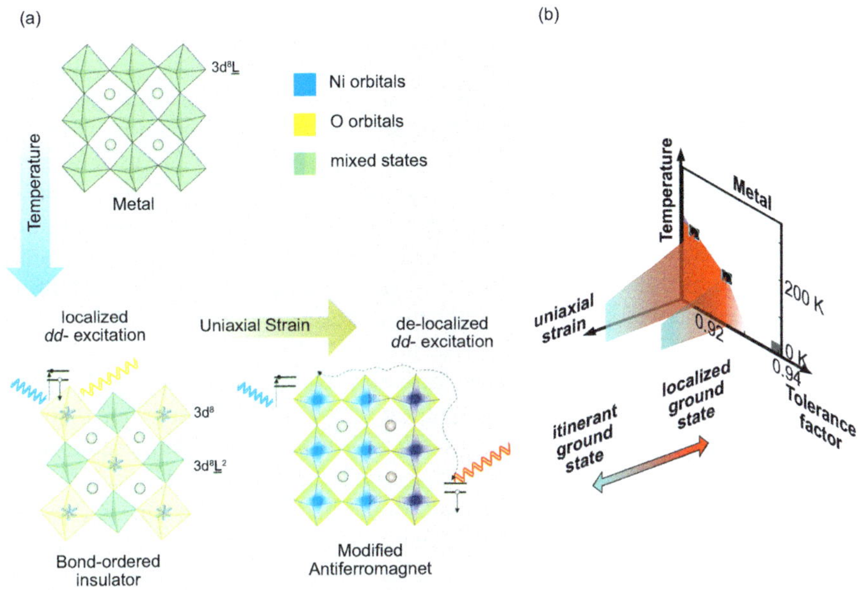

Fig. 3.46 Key understanding about effect of helium implantation on ncikelates: **a** The high temperature metallic state hosts highly hybridized $3d^8\underline{L}$ state which on cooling separates into $3d^8$ and $3d^8\underline{L}^2$ (right) fixed-q energy scan on the magnetic peak, described by sit-selective Mott transition. The dd-excitation from resonant L_3 edge is created and collapsed on the same octahedron to emit the X-rays detected. Under strain this ground state becomes more itinerant specially in the d-orbitals while bond order and Ni-O hybridization decrease. Hence, we conjecture a itinerant blue could of d-electrons such that dd-excitation is created and collapsed on different octahedra. **b** Thus, this strain works orthogonal to the typically explained phase diagram of nickelates depicted for the region investigated. The ground state becomes more itinerant

prior reports of metallic antiferromagnetism under compressive strain in heterostructures. A Landau theory based on quarter-filled Ni-e_g bands captures key features of this regime. Helium implantation is a unique manipulation tool which tunes the layer separation and allows direct electronic state manipulation of quantum materials. We are barely scratching the surface of the incredible tuning property this method allows. Interesting studies can be taken up in materials with different structural coupling between layers such as van der Waals materials. Strongly interacting localized physics could be turned into weakly interacting itinerant physics offering plethora of possible avenues in different quantum materials.

Finally, with the method understood we turn towards the primary goal of this chapter—creating antiferromagnetic pattern and observing them using coherent resonant scattering. The reason for this detour is justified in terms of understanding the changes we were to be expecting when we focus the implantation down.

3.6 Results: Pattern Formation Using Focused Ion Beam

Once we have realized the robust effect we bring into the nickelates uniformly implanted with He ions we move to the ultimate challenge of creating mesoscale textures of electronic states using a focused ion beam of helium. We used a helium ion microscope to implant the helium onto the substrate. We will show that the focused ion beam successfully tunes the local structure of the substrate and the uniaxial strain remains localized to the thin film in the implanted region only. Since the uniaxial strain changes the electronic character and the phase transition of the nickelate film, we will see by local implantation of helium ions we will be able to create artificial mesoscale electronic structures.

We used a dose of 30e14 ions/cm^2 for both $PrNiO_3$ and $NdNiO_3$ film. For the $PrNiO_3$ film, this dose completely suppressed the transition down to liquid helium temperature, but for the $NdNiO_3$ sample, it drives the transition temperature below 90 K. The only difference we have from the uniformly implanted case is that the energy of the helium ion beam is limited to 25 keV from the helium microscope specification at QB3 of UC Berkeley. According to our previous SRIM simulation, this is sufficient energy for the ions to pass the thin film and get implanted into the substrate. Thus we do not expect any different behavior.

We first discuss the pattern we created on the $PrNiO_3$ sample. For any investigation related to a mesoscopic pattern, we need to help with creating a map of the sample. Often these are done through several marks which are called fiducials. During the X-ray experiments, they help us find the tens of micron-scale patterns in a millimeter scale thin film. Thus, after growth and characterizing we created fiducial marks on the sample depending on the experiment at hand. For the $PrNiO_3$ sample, we investigated the success of electronic pattern creation using X-ray photo-electron emission spectroscopy (X-PEEM). Since, as per our discussion above, even the room-temperature state is different in electronic configuration, we utilize that to create contrast in the photoemission signal.

3.6.1 Resonant X-Ray Photo Electron Emission Microscopy

Resonant photoemission ejects a core electron from an occupied level. Ejection of the core electron creates a core hole which is further quenched by an electron from the continuum or an electron from the higher valence shells. If the core hole is filled with an electron from a higher valence shell another electron is ejected in the process. Such secondary electron emission is known as the Auger effect. All of these electrons come out of the surface of the sample exposed to the resonant radiation. The total electron yield can be parallelly measured connecting a ground wire from the sample through an ammeter. This current is recorded in X-ray absorption spectroscopy experiment. Thus, in resonant photoemission, we focus directly on the electrons instead of the replenishing current. However, they are directly related.

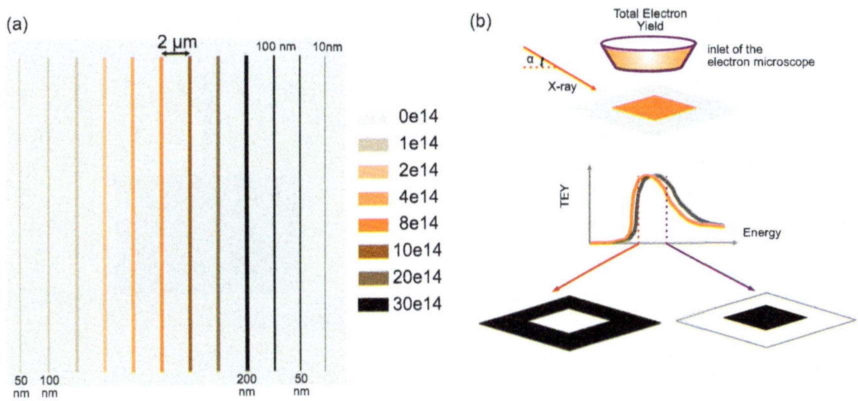

Fig. 3.47 X-ray photo-electron emission microscopy: **a** Sample design **b** Schematic of contrast mechanism: the orange TEY signal comes from the implanted electronically modulated area. Thus, sitting at two different energies creates the contrast to spatially resolve the electronic textures

The electrons have a small mean free path in the sample. Thus, only the electrons very close to the surface ~2–10 nm make it out of the surface for detection. In the microscopy mode of this experiment, we set an external electric field perpendicular to the surface of the sample. For our experiment, it was set at 18 keV. Right after ejection, the electrons are subjected to this field and accelerated towards the objective lenses. At the focus of the objective lenses, a two-dimensional electron detector is placed to record the distribution of the electrons in real space. In this way, we get an image of the electron-emitting parts of the sample under focus for a given energy. The brightness of the pixel is directly related to the count of electrons leaving that position on the sample for a given energy.

Contrast Mechanism: A typical X-PEEM experiment relies on finding the contrast of the sample surface as a function of energy. This is schematically shown in the figure. To design a contrast, we first study the TEY of the XAS on the sample and focus on the energies where they deviate in intensity. For our sample, we know that near peak A the intensity in TEY is slightly higher for helium-implanted samples, which can be utilized to create the contrast. At this energy, electrons emitted from the regions with different electronic configurations due to the helium implantation will be higher in number than in a different position in energy where the pristine ejects more electrons. Thus, if we take an image with incoming energy near peak A the implanted regions will appear brighter. We can further increase the contrast by utilizing images taken at two energies. If we divide the image taken at the first energy by the image taken at the second energy, the image will appear brighter where the electronic configuration is changed due to helium implantation.

Sample preparation and Mounting: We carried out the XPEEM experiments in beamline 11.0.1.1 of the Advanced Light Source of the Lawrence Berkeley National Lab. The lowest temperature we could achieve was 150 K and hence we explored the electronic phase heterogeneity in the metallic phase on samples where we developed

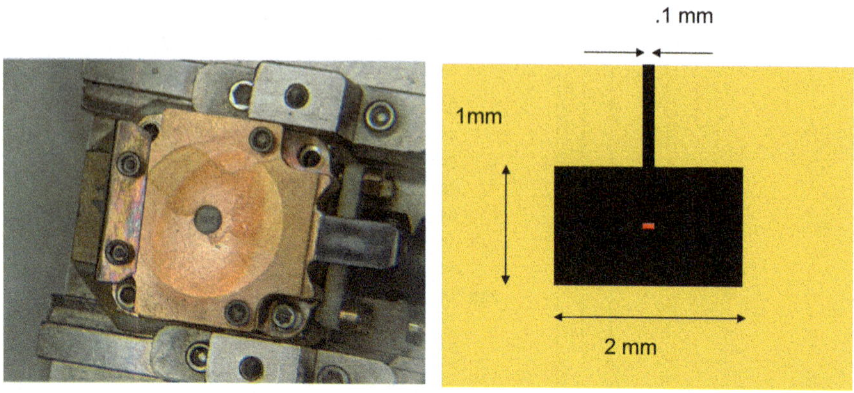

Fig. 3.48 Sample design and sample mount the central red box contains all the pattern: **a** The sample in the X-PEEM holder where the gold-fiducial on the sample can be seen. **b** The patterned area in red with respect to the fiducial schematically presented

line patterns of different thicknesses and helium doses. The schematic of the sample is shown in the Fig. The highest dose we explored is 30e14 ions/cm^2 similar to the uniform helium implantation. Each line was separated from the other by 2 μm. In the central region, we wrote the pattern with an equal thickness of 200 nm but varying helium dose. Both ends, on the other hand, had lines with varying thickness for damages 30E14 ions/cm^2 (200 nm, 100 nm, 50 nm, and 10 nm), and 1e14 ions/cm^2 (200 nm, 100 nm, and 50 nm). From the uniform helium implantation, we know that at this dose the film remains metallic down to the lowest temperature. The objective of this experiment was to first confirm that we can write electronically heterogeneous patterns with a focused helium ion beam and then characterize the patterns written (Fig. 3.48).

The samples were prepared by our collaborators from the Dynes lab of the University of California San Diego and developed in the helium microscope of the University of California Riverside. Helium dose was varied from 1e14 ions/cm^2 to 30e14 ions/cm^2 in multiple steps to check if the pattern forms similarly among different doses. We also varied the thickness of the lines to check if it affects the quality of the patterns drawn. The helium beam was focused down to 0.5 nm and the resolution of the XPEEM experiment was 30nm. We further developed a fiducial pattern which is shown in the Fig along with the sample mounting set up in the XPEEM chamber. It can be seen that the sample sits under a Cu cover. This puts a design consideration where the pattern to be investigated must be drawn at the center of the sample away from the Cu rim. The fiducial design was developed by depositing 40 nm Ti/Au with a custom-designed mask. The empty channel in the gold deposition is intended as a guiding direction to find the pattern at the center. The channel thickness was kept at 100 μm similar to the length of the pattern to ease the search process.

Exposure control: One main consideration while doing an XPEEM experiment is to control the exposure. Extremely high brilliance beam is focused to a small

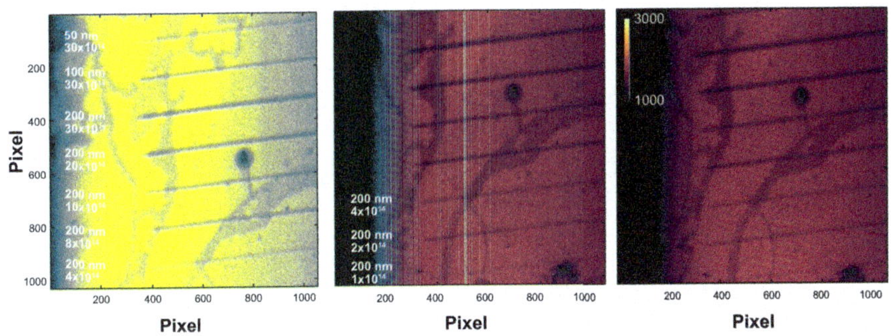

Fig. 3.49 XPEEM images showing features and beam damage with exposure: (left to right) subsequent images taken with time shows dimming of the intensity due to exposure. The colorbar for all of the panels are shown on the right-most panel. We can clearly see most of the lines with higher dose and thicker lines expectedly offering most contrast

micron size spot increasing the resonant flux density extremely high. Resonant X-rays tune to the valence electrons and can damage the film for such a high flux density. Thus, at the beginning of the experiment, we found a region to test our exposure and find the optimum exposure such that the sample electronic environment is not altered while we have sufficient count to discern 30 nm scale features, which is the instrument resolution. We found that for our film both pristine and helium-implanted region are damaged with long exposure. By trial and error, we found that a slit opening of $80 \times 500\,\mu$m with 0.5 s exposure was optimum. To increase the statistics, we repeated each scan 8 times at a given input energy. The beam was incident on the sample at an angle of 45° and an Au foil was inserted in the beam path to further decrease the flux. To make sure we have not damaged the sample while collecting data, we collected a reference image before and after the spectro-microscopy scan. In the region of interest, we collected more images after data collection to show the effect of X-ray radiation. From left to right, the exposure time increases and the photoelectron count drops. All panels are scaled to the same colormap to highlight this feature. Furthermore, we realized that at low temperatures, the beam damages are better sustained by the sample. Thus, we cooled the sample down to $\sim$160 K while collecting data (Fig. 3.50).

Result: The Fig. 3.49 provides the first evidence of our ability to write localized patterns of a different electronic state. The sample is not etched, but only the substrate underneath is implanted with helium. We observe that the pattern written with a helium dose of 30e14 ions/cm^2 is the most prominent. The image is taken with the Ni-L3 resonance of energy, which we had been referring to as peak B. Thus, the implanted regions appear dark. We have taken multiple cuts along the perpendicular direction to the elongation of the stripe. We averaged the cuts to smooth out the fluctuations and extract the width of each artificially patterned electronic feature. The extracted data and the intended design are shown in the Table 3.9.

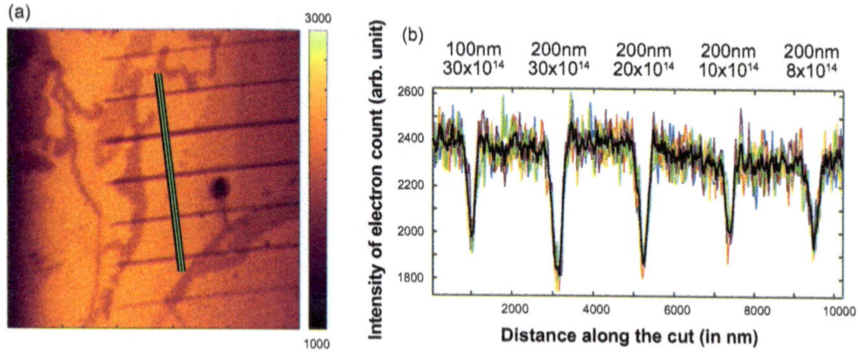

Fig. 3.50 Taking cuts on the X-PEEM images and verifying widths: **a** We show the lines we used to take the cut vs intensity plots on the right marked by green and black solid lines. **b** Each cuts corresponding to the six lines of panel (**a**) are shown and then averaged to improve signal to noise ratio. The average cut is represented by the solid black line. Above each line intended thickness and the dose in ions/cm^2 is mentioned

Table 3.9 Extracted width compared with intended design of the artificial electronic pattern

Dose	Intended thickness (nm)	Extracted thickness (nm)
8e14	200	210
10e14	200	210
20e14	200	225
30e14	200	255
30e14	100	180

We extracted the width of each line from the full width at half-max of the intensity profile. It is worth noting that the intensity from the helium-implanted regions increases with a lower dose thereby decreasing the contrast. To understand this, we conduct a spectroscopy experiment with resolution of 28 nm. The intention was to compare the total electron yield with the uniformly implanted sample with a localized implanted region. We want to ensure that the region implanted with 30e14 ions/cm^2 behaves electronically similar to the sample we studied in the previous section (Fig. 3.51).

Spectromicroscopy at 150 K

We have, so far, shown that the focused helium implanted region is electronically different from the pristine region. The continually decreasing trend of the contrast from the cuts in the previous figure is a first hint that they are different from each other as well. However, we want to confirm if the continuous tunability of the electronic phase observed during the uniform helium implantation still survives in the focused implantation limit. We compared the XAS TEY measured on the uniformly implanted samples with the total count from a well-defined region as we ramped the energy of the

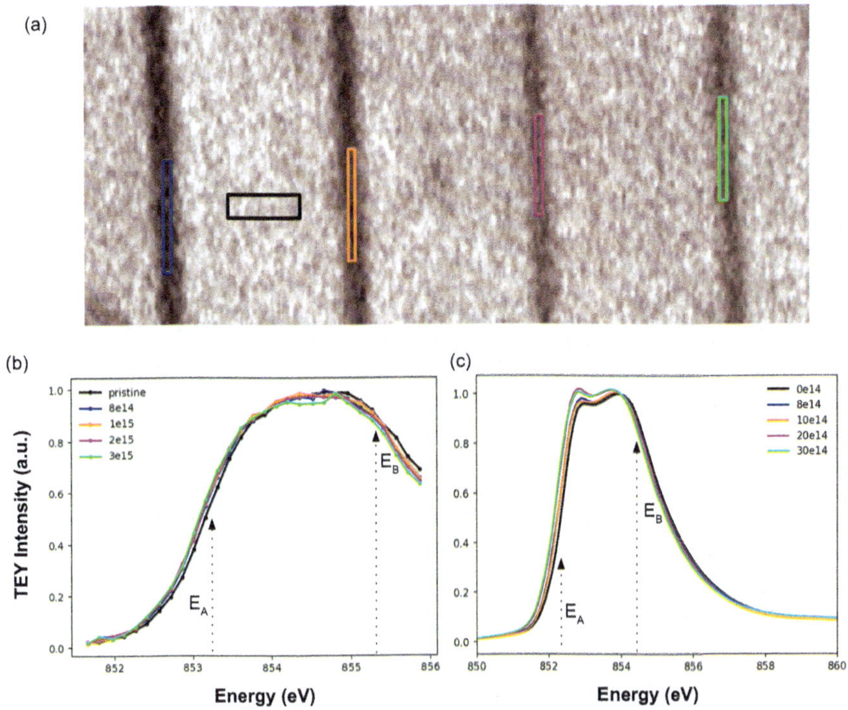

Fig. 3.51 X-ray Absorption characterization confirming electronic phase tuning: **a** Collecting spectra from the focused helium implanted region marked by colored boxes corresponding to the same line color XAS extracted shown in (**b**). Sample temperature was 150 K. **c** Shows XAS measured on uniformly implanted samples at this temperature of 150 K. The intensity evolution at E_A and E_B shows changes the electronic character similar to the uniformly implanted samples establishing the equivalence of electronic tuning between uniform and focused ion implantation

incoming beam. Thus, they should be directly comparable. Furthermore, we set the polarization to be vertical, and with 45° incidence angle we have an equal contribution from the in-plane and out-of-plane components. For the tey measured in the uniformly implanted samples we have also averaged the polarization for direct comparison. Since the energy resolution in XPEEM is poorer than the energy resolution of the XAS beamline we only see a qualitative match. This is shown in the Fig below where we have shown side by side the XAS collected from the $PrNiO_3$ samples implanted with uniform dose of 0, 8, 10, 20, and 30e14 ions/cm^2 with the XAS extracted from defining a region of interest as the small boxes and averaging the count in the boxes for each energy. We see an increase (decrease) in weight for the states near the E_A (E_B) energy with increasing helium dose for both uniformly and focused helium implantation. This result proves that the focused helium implantation changes the electronic configuration, similar to a uniform helium implantation. Thus, the suppression of the phase transition will be observed in this localized region. This is precisely what we wished for. If we lower the temperature on this sample, the helium-

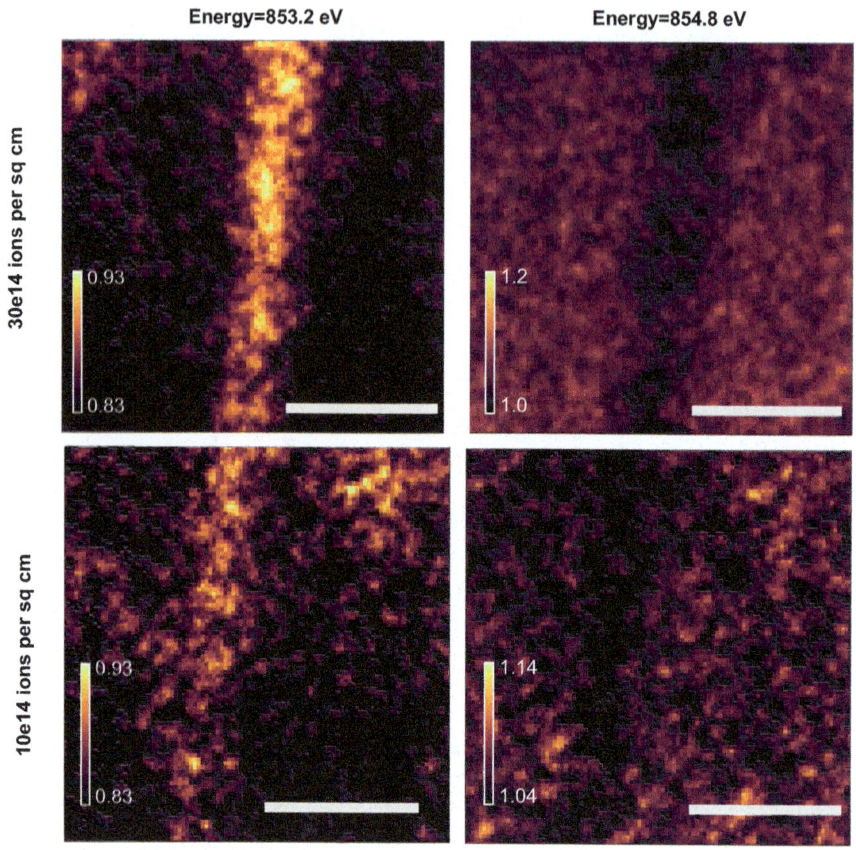

Fig. 3.52 Spectromicroscopy on the helium implanted lines: Spectromicroscopy observed on the 30e14 and 1e14 ions/cm^2 implanted 200 nm wide lines showing a greater spread but better uniformity for the 30e14 ions/cm^2 lines. The local inhomogeneity remains identical after helium implantation even though the energy to which the region responds (brighten up) changes. The scale bar represents 500 nm

implanted region will not turn into an antiferromagnetic insulator whereas the pristine region will become antiferromagnetic creating an artificial lateral antiferromagnetic texture.

Before we move onto that final step of this project of creating and verifying antiferromagnetic texture, we note the heterogeneity of the pulsed laser deposited samples which can be seen in the spectromicroscopy images taken at the B-peak in the Ni-L3 resonance profile. That heterogeneity survives inside the implanted region indicating that the individual grains of the thin film remain undisturbed by helium dose as high as 30e14 ions/cm^2. From the width evaluation, we found that for the dose of 30e14 ions/cm^2 may the feature spreads more than the intended dimension, however, that dose provides us with the cleanest line profile as seen from the comparative Fig. 3.52. This is also evidenced by the smooth drop of the line profile in Fig. 3.47.

3.6.2 Resonant Coherent Soft X-Ray Scattering

We have confirmed so far that with helium implantation we can locally tune the electronic configuration of the nickelates. The room-temperature electronic state was found to be similar between the locally tuned and uniformly implanted samples. Since the electronic state at high temperature was similar to the uniformly implanted sample, we expect that it will undergo the coupled phase transition similarly. Thus, for the selected implantation dose of 30e14 ions/cm^2 the helium implanted $PrNiO_3$ film should not undergo the coupled phase transition down to liquid helium temperatures, whereas, the implanted $NdNiO_3$ film should show an antiferromagnetic transition around 90 K. Thus, if we set the temperature above 100 K we expect to observe implanted regions to remain paramagnetic while the pristine region will turn into antiferromagnet.

The pattern we created along with the sample fabrication details is shown in the Fig. 3.53. The intention was to implant the lines of helium equally spaced over a region on the order of 20 μm. We patterned two regions: one with 500 nm wide implanted regions separated by 200 nm pristine region, and the other with 1000 nm wide implanted regions separated by 200 nm pristine region. We used a dose of 30e14 ions/cm^2. We wanted to create an antiferromagnetic grating which will leave the signature of its periodicity on the reciprocal space reflection on the magnetic peak as a small-q modulation. The spatial parameters were motivated from forward modeling the scattering outcome discussed below. With these considerations, we set out to fabricate the focused ion implanted samples.

Sample fabrication: As deposited thin film samples were used to developed patterns at helium microscope of Quantitative Biosciences at UC Berkeley (QB3-Berkeley). The implantation energy was 30 keV. From our evaluation this is sufficient for the ions to pass through the thin film and get implanted into the substrate about 100 nm below the interface. Such high-energy helium was focused on the sample plane using the electro-magnetic lenses. The beam current was set at 3.36 pA with a dwell time of 1 μs which deposited 30e14 ions/cm^2 in a region. Before taking the sample to the helium facility we deposited gold fiducial according to the experiment at hand. Unlike the XPEEM experiment, nano-diffraction experiment usually demands creation of microstructures to be investigated near the edges of the sample because of the limitation from zone plate optics.

Fresnel zone plate is used to focus the beam on the sample for nano-diffraction measurements, which is discussed in detail in Chap. 6. However, for the present discussion, we note that an order sorting aperture is required for the zone plate and is placed between the zone plate and the sample. For example, at CSX-23 ID beamline, where we investigated the helium-implanted patterns. The approximate separation between the zone plate and the OSA was approximately 8.6 mm for a focal length

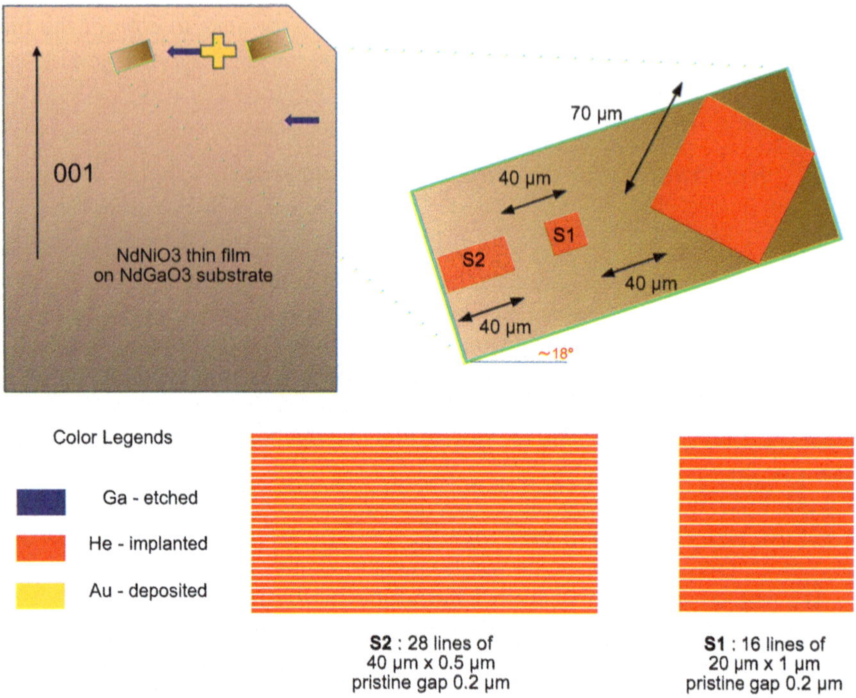

Fig. 3.53 Sample fabrication for resonant nano-diffraction: focused helium beam implanted $NdNiO_3$ thin film with grating like pattern S1 and S2. Detailed dimensions are shown in the figure along with the fiducial marks. The pattern is rotated by 18° with respect the edge to have the grating modulation along perpendicular direction to scattering plane. The Ga etches are to acts as a fiducial as they will appear dark at Ni resonance

of the zone plate at 10.3114 mm with a depth of focus of 6 μm at 852 eV. Thus, the sample to OSA separatio is only about 1.72 mm. Thus, if we have a sample that requires the zone plate-OSA assembly to move 2 mm downstream to focus on a point 2 mm inside the edge of the sample, then the assembly would crash. This mechanical limitation motivates us to design patterns near the edge of the sample so that we have sufficient room for the zone plate assembly (Fig. 3.54).

In nano-diffraction measurements it is useful to write different scales of fiducial from the microscopic length scale to the length scale of intended feature so that at every step of zooming in we have fiducial marks to act as guides. This is what we describe here. We deposited a gold cross of ~250 μm followed by a Ga-etched arrow of similar dimension. We later realized during the experiment that the Ga etching was not required. Then we implanted a region of $70 \times 70\ \mu m^2$ on both sides of the gold cross. Right next to the helium-implanted region we drew the patterns. Each pattern contains one region of 16 lines of implanted regions with dimensions of 1 μm×20 μm with 200 nm spacing in between (S1), and 28 lines of implanted regions with dimensions of 0.5 μm×40 μm with 200 nm spacing in between (S2).

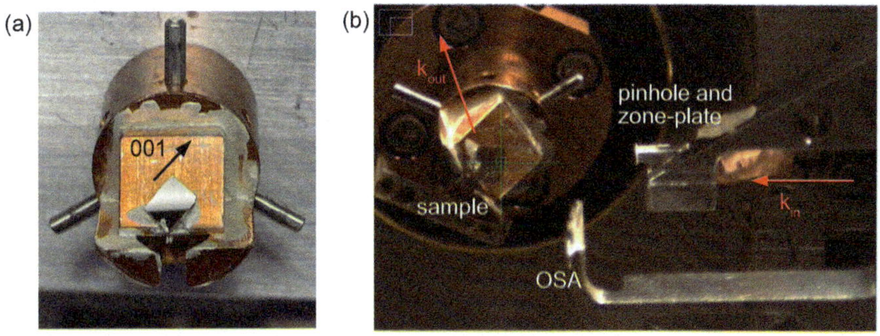

Fig. 3.54 Experimental set up and mounting of the sample: a We mounted the sample rotated by 45° in ϕ and on a wedge to provide the χ tilt of 54° **b** The beam propagation direction along with the optics assembly used for this experiment. The optics assembly can move along the beam or perpendicular to the beam. During nano-diffraction we scanned the zone plane and OSA together in the plane perpendicular to the beam propagation

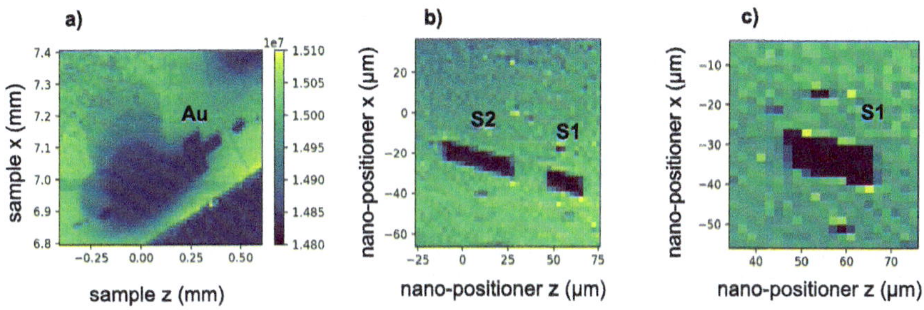

Fig. 3.55 Finding the pattern through guidance by the fiducial: a–c We find a feature at a scale and zoom in to find the next feature on that scale till we reached S1. This is typical strategy for nano-diffraction experiment where we create different guiding fiducials at different scales

The implanted regions will remain paramagnetic below the transition with the pristine regions becoming antiferromagnetic. Thus S1 and S2 will have 200 nm wide line of antiferromagnetic regions separated by 1000 nm and 500 nm of paramagnetic regions, respectively. Thus, if we see the modulation in the scattered intensity due to the antiferromagnetic grating the modulation will have different spatial frequency. We further developed M1 to M4 which are implanted regions of 4×4 μm^2 area with a pristine square hole of 100–500 nm side-length at the center. However, due to time limitations, we could not finish our investigation on them (Fig. 3.55).

Coherent Resonant Soft X-Ray Scattering

The samples were created right before the experiment date and was not heated or treated by anything other than acetone coated wipe to clean the surface. The sample was then mounted on a wedge similar to the experimental conditions described in this chapter. With the sample mounted like that we set the polarization to π and energy to the Ni-L3 edge at 853 eV. All other experimental conditions pretty much resembled the experiment described in the previous chapter with only one crucial difference: we used Fresnel zone plate with OSA and not a pinhole. We lowered the temperature to 112 K where the pristine sample showed a clear AFM peak at $(\frac{1}{2}, 0, \frac{1}{2})$. We know from our investigation of the uniformly implanted sample that at this temperature the implanted $NdNiO_3$ must remain metallic-paramagnet. We killed the feedback correction to the diffractometer motors after aligning the diffractometer properly to the antiferromagnetic peak so that it does not influence our positioning. For the rest of the experiment we moved the sample position or rastered the beam using nano-positioners which can move the optics over the beam to move the focus on the sample.

Since the depth of the focus of the zone plate is roughly 6 μm at the diffraction condition of $\theta = 55°$, $\phi = 45°$, and $\chi = 54°$ (where the last two angular motion were provided by how we mount the sample on a wedge) it is very easy to become out of focus if we are traversing the beam on the sample. Thus, it is important to fix the distance of the zone plate-OSA assembly by rastering over a known feature on the sample and making sure that the contrast of the feature is the best for the selected distance among many other choices. This optimization of the zone plate is an added condition for zone plate-based coherent scattering. For our experiment, we checked the contrast on the antiferromagnetic gratings (S1 and S2) to be investigated to fix the distance of the zone plate-OSA assembly. We conducted the experiment in two different modes both effectively showing the same information. First we conducted a nano-diffraction experiment which is a scanning mode experiment to investigate the texture. At the diffraction condition set to the AFM peak we scanned the beam on the sample using nano-positioners. Here we measured the total scattered intensity for every point the beam landed on the sample. We scanned with a resolution of 100 nm steps with the a focus size where the contrast of the features were maximum. The nano-diffraction measurements on the patterns S1 and S2 clearly showed the beam spot convoluted intensity modulation due to the grating pattern we intended to create. This is a clear proof that at low temperature the implanted areas remain paramagnetic but the pristine areas become antiferomagnetic. From our measurement, we can also find the separation between the pristine regions and see that it is approximately double for the S1 pattern as expected (Fig. 3.56).

Another way we find the signature of the AFM grating is on the single shot scattered intensity recorded on the detector. This is a remnant of the discussion we had in this chapter where meso-scale dilute patterns leave its mark on the intensity modulation on the Bragg peak. When the zone plate position is not optimal, the beamspot is not tightly focused on the sample and we have a broad beam covering multiples of the fringes and we see a clear supermodulation on the scattered intensity as shown in the figure below. The frequency of modulation(Δq) is inversely related to the separation of two antiferromagnetic strips. Since S1 has twice the separation in real space than S2, the separation of fringes on the AFM Bragg peak is lower by a factor of 2 (Fig. 3.57).

Thus, finally, we demonstrate the creation of an antiferromagnetic (AFM) grating structure in a nickelate thin film using a local post-growth method of focused helium ion implantation.

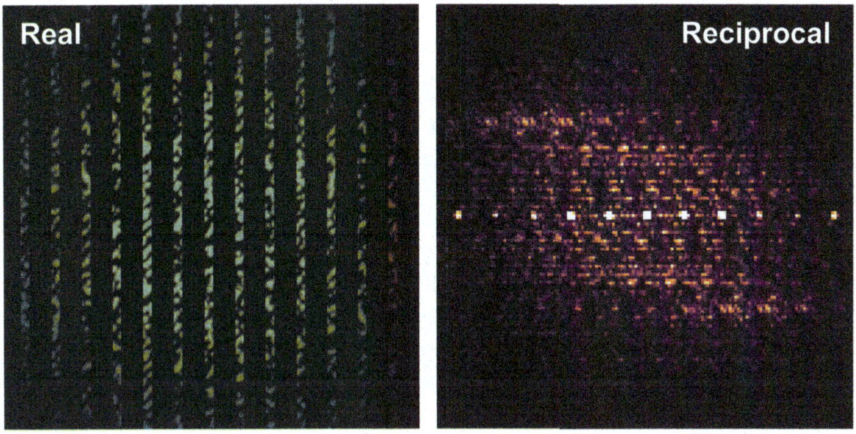

Fig. 3.56 Modelling of the experiment and expectation from the grating pattern for a defocused coherent illumination: (On the right) The reciprocal space pattern is just the amplitude Fourier transform of the (left) real space pattern. We see super-modulation due to the artificial grating structure imposed on the natural antiferromagnetic domain formation

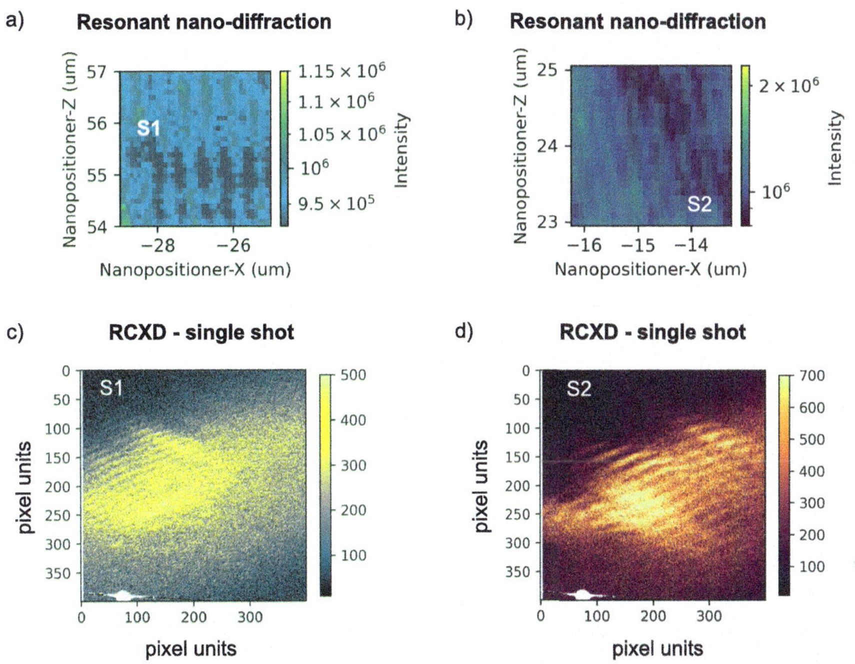

Fig. 3.57 Nano diffraction and resonant coherent X-ray diffraction measurement: **a**, **b** Resonant nano-diffraction showing ~1200 nm period on S1 and ~750 nm on S2 as expected. **c**, **d** Single snapshot of the detector during Resonant Coherent X-ray Diffraction on the fabricated AFM grating shows interference pattern on the AFM peak. The fringe spacing decreases for S1 compared to S2 in-line with increased line-spacing in real space

3.7 Conclusion

In this chapter, we achieved tuning electronic character of a strongly correlated electronic system through strong-electron lattice coupling. We modified the lattice, through which we changed the electronic character by helium implantation. We showed here that implantation of helium ion into the substrates changes the uniaxial out of plane strain for the epitaxial film while keeping the in-plane biaxial strain unperturbed. We also observed that the helium implantation did not change the electron-lattice coupling and we could increase the itinerant character of the d-electron by tuning the strain along the out-of-plane direction only. We monitored the change in bond order, charge order, transport, and magnetic order of the sample.

The magnetic order revealed a change in the character of the magnetism. Following the change in the localized character and decrease in the Bond order, we believe we gradually changed the electronic environment for the d-electrons from a *site selective Mott transition* to *Fermi surface guided density wave transition*. This is continually tunable and proportional to the helium dose we offer. In this way we have shown a controlled mechanism of strain tuning on rare-earth nickelates and tuned the nature of electronic correlation. Such a degree of control in tunability post-growth of epitaxial nickelate can be used to further expand the phase diagram like we did here. For example, we can employ the same techniques on oxygen-deficient nickelates or hole-doped nickelates to check for the effect. From our X-ray absorption spectroscopy measurement we showed that the high-temperature electronic state was also modified in this process. Thus we have shown that we can change the correlation in a bad metal state by the uniaxial strain tuning, which can be further employed other relevant materials.

We can further follow up on the project by studying the fluctuations in the implanted samples by X-ray photon correlation spectroscopy: How the fluctuation changes as we increase the helium implantation, and how it relates to electronic phase, can be an important study characterizing the dynamic charge correlation in this material. In [21], the authors numerically found significant evidence for dynamic charge correlation, which can influence the magnetic energy response. We observe the magnetic energy response to change as a function of the helium dose. Thus, to complete the story a follow-up work can correlate and highlight the relevance of dynamic charge correlation measuring it directly.

Furthermore, taking advantage of helium microscope this strain can be tuned locally to create lateral heterostructures. Utilizing experimental methods sensitive to local electronic states, we were able to image the lateral heterostructures that we artificially created. We have, effectively, created a metastable phase of a complex material. Since the implanted region is highly itinerant it will be a curious investigation to follow the effect of two implanted lines on the pristine layer between them. We can study the fluctuation on an ordered peak such as the magnetic Bragg peak measured on a patterned region. This will be an interesting problem in mesoscopic physics of such complex materials. With promise of so many interesting physics questions around, with the new physical system we have created, we conclude this section.

References

1. Middey S, Chakhalian J, Mahadevan P, Freeland JW, Millis AJ, Sarma DD (2016) Physics of Ultrathin films and heterostructures of rare-earth nickelates. Ann Rev Mat Res 46:305–334. Publisher: Annual Reviews
2. Guo H, Dong S, Rack PD, Budai JD, Beekman C, Gai Z, Siemons W, Gonzalez CM, Timilsina R, Wong AT, Herklotz A, Snijders PC, Dagotto E, Ward TZ (2015) Strain doping: reversible single-axis control of a complex oxide lattice via helium implantation. Phys Rev Lett 114:256801
3. Toulouse C, Fischer J, Farokhipoor S, Yedra L, Carlà F, Jarnac A, Elkaim E, Fertey P (2021) Patterning enhanced tetragonality in BiFeO3 thin films with effective negative pressure by helium implantation. Phys Rev Mat
4. Allen FI (2021) A review of defect engineering, ion implantation, and nanofabrication using the helium ion microscope. Beilstein J Nanotechnol 12:633–664
5. Herklotz A, Wong AT, Meyer T, Biegalski MD, Lee HN, Ward TZ (2016) Controlling octahedral rotations in a perovskite via strain doping. Sci Rep 6(1):26491. Publisher: Nature Publishing Group
6. Sharma V, Herklotz A, Ward TZ, Reboredo FA (2017) Designing functionality in perovskite thin films using ion implantation techniques: assessment and insights from first-principles calculations. Sci Rep 7(1):11166. Publisher: Nature Publishing Group
7. Bisogni V, Catalano S, Green RJ, Gibert M, Scherwitzl R, Huang Y, Strocov VN, Zubko P, Balandeh S, Triscone J-M, Sawatzky G, Schmitt T (2016) Ground-state oxygen holes and the metal–insulator transition in the negative charge-transfer rare-earth nickelates. Nat Commun 7(1):13017. Publisher: Nature Publishing Group
8. Park H, Millis AJ, Marianetti CA (2012) Site-selective mott transition in rare-earth-element nickelates. Phys Rev Lett 109(15):156402
9. Lau B, Millis AJ (2013) Theory of the magnetic and metal-insulator transitions in r nio 3 bulk and layered structures. Phys Rev Lett 110(12):126404
10. Georgescu AB, Millis AJ (2022) Quantifying the role of the lattice in metal–insulator phase transitions. Commun Phys 5(1):135
11. Peil OE, Hampel A, Ederer C, Georges A (2019) Mechanism and control parameters of the coupled structural and metal-insulator transition in nickelates. Phys Rev B 99(24):245127
12. Hepting M, Minola M, Frano A, Cristiani G, Logvenov G, Schierle E, Wu M, Bluschke M, Weschke E, Habermeier H-U et al (2014) Tunable charge and spin order in prnio 3 thin films and superlattices. Phys Rev Lett 113(22):227206
13. Frano A, Schierle E, Haverkort MW, Lu Y, Wu M, Blanco-Canosa S, Nwankwo U, Boris AV, Wochner P, Cristiani G et al (2013) Orbital control of noncollinear magnetic order in nickel oxide heterostructures. Phys Rev Lett 111(10):106804
14. SungBin L, Chen R, Balents L (2011) Landau theory of charge and spin ordering in the nickelates. Phys Rev Lett 106(1):016405
15. SungBin L, Chen R, Balents L (2011) Metal-insulator transition in a two-band model for the perovskite nickelates. Phys Rev B 84(16):165119
16. Kim J-W, Choi Y, Middey S, Meyers D, Chakhalian J, Shafer P, Park H, Ryan PJ (2020) Direct evidence of the competing nature between electronic and lattice breathing order in rare-earth nickelates. Phys Rev Lett 124(12):127601
17. Lu Y, Frano A, Bluschke M, Hepting M, Macke S, Strempfer J, Wochner P, Cristiani G, Logvenov G, Habermeier H-U, Haverkort MW, Keimer B, Benckiser E (2016) Quantitative

determination of bond order and lattice distortions in nickel oxide heterostructures by resonant x-ray scattering. Phys Rev B 93(16):165121 April
18. Georgescu AB, Millis AJ (2022) Quantifying the role of the lattice in metal–insulator phase transitions. Commun Phys 5(1):1–14. Publisher: Nature Publishing Group
19. Freeland JW, van Veenendaal M, Chakhalian J (2016) Evolution of electronic structure across the rare-earth RNiO3 series. J Electron Spectrosc Relat Phenom 208:56–62 April
20. Disa AS, Georgescu AB, Hart JL, Kumah DP, Shafer P, Arenholz E, Arena DA, Ismail-Beigi S, Taheri ML, Walker FJ, Ahn CH (2017) Control of hidden ground-state order in NdNi O 3 superlattices. Phys Rev Mat 1(2):024410 July
21. Green RJ, Haverkort MW, Sawatzky GA (2016) Bond disproportionation and dynamical charge fluctuations in the perovskite rare-earth nickelates. Phys Rev B 94:195127 Nov
22. SungBin L, Chen R, Balents L (2011) Landau theory of charge and spin ordering in the nickelates. Phys Rev Lett 106(1):016405

Chapter 4
Phase Competition With Antiferromagnetic Order In Fe Intercalated NbS_2

Abstract After imaging dilute antiferromagnetic domains, and fabricating antiferromagnetic pattern, here we discuss competition between two antiferromagnetic phases in Fe-intercalated NbS_2. First we introduce phenomenology in this highly tunable magnetic material. Then we highlight our discovery of charge order in this material emergent from magneto-elastic coupling. This observation becomes useful in understanding our main observation of this chapter where we study the phase competition phenomena using coherent resonant soft X-ray scattering. Detailing the experimental nuances we layout the observation first. Next, we discuss the data analysis rooted in extracting cross-correlation of the speckle pattern. Finally, aided by the observations we introduce a theoretical model which nicely captures the key observations in the McMillan-Ginzburg-Landau framework.

4.1 Introduction: Coherent Scattering to Study Phase Competition

Phase competition and phase coexistence are ubiquitous in strongly correlated electronic systems. When multiple electronic degrees of freedom such as spin, orbital, charge, and lattice couple to each other, a spatially inhomogeneous ground state often emerges. Such textured material unsurprisingly deviates from linear behavior and small perturbations are met with large responses. High-temperature superconductors are one such example which remains enigmatic to this day. Small tuning of the electron or hole doping percentage completely changes the electronic character of the material. Moreover, the superconducting state emerges competing and often coexisting with a charge density wave state. Thus, understanding of the phase competition and phase coexistence is attributed to the mystery of understanding the high-temperature superconducting phase of cuprates.

To understand phase competition and phase coexistence we need tools to characterize electronic heterogeneity and its evolution under perturbation. Coherent resonant scattering perfectly fits the requirement. On one hand, the resonance condition makes it sensitive to the electronic states, and on the other coherence makes it sensitive to the real space textures. Here we will explore the potential of coherent resonant

R. Basak, *Nucleation, Fabrication, and Competition in Antiferromagnetic Textures Studied with Coherent Resonant Soft X-ray Scattering*, Springer Theses,
https://doi.org/10.1007/978-3-032-19199-1_4

scattering to understand phase competition in a highly tunable magnetic material: Fe intercalated NbS_2.

We first discuss previous works that have inspired us to conceive the project: starting with the strange phenomenology of resistance switching observed in this material and subsequent in-depth study of highly tunable magnetic ground state explored by neutron scattering. . Then, we discuss our discovery of a concomitant charge order state that couples to an external magnetic field providing us with sufficient evidence for a strongly coupled electronic environment. Finally, we characterize phase competition in such a system using coherent resonant soft X-ray scattering.

4.2 Phenomenology of $\mathbf{Fe_xNbS_2}$

$Fe_{\frac{1}{3}+\delta}NbS_2$ is formed by intercalating transition metal ions of Fe into van der Waals bonded layers of NbS_2. NbS_2 belongs to the large class of transition metal di-chalcogenides (TMDC, TA_2) with $T = Nb, Ta, V, Mo$ and $A = S, Se$. Two-dimensional TMDCs host coupling of electronic degrees of freedom making them attractive not only for studies of basic physical phenomena but also for technological applications. These materials are a prototypical example of the charge density wave phase emerging from the Fermi surface instabilities of weakly interacting itinerant electrons at low temperatures. They also host phonon-mediated superconductivity at lower temperatures. However, NbS_2 stands out of the pool because it is the only TMDC to host superconductivity without a charge density wave state indicating a different electron correlation environment.

Structure

Including Fe atoms in between chalcogen layers increases the electronic correlation in this system. For TMDCs there are two thermodynamically stable phases: 2H or trigonal prismatic and 1T or distorted octahedral, categorized following the stacking of transition metal and chalcogen atoms (see Fig. 4.1). For stoichiometric intercalation of $x = \frac{1}{3}$ in the formula unit of Fe_xNbS_2 the Fe atoms arrange into a stacked bilayer arrangement of $\sqrt{3} \times \sqrt{3}$ super-lattice in between layers of 2H-NbS_2. Such an arrangement breaks the inversion symmetry of the crystal unit cell, which results in a spin splitting of the electronic bands driven by spin-orbit interaction.

Thermodynamic Characterization

First, the intercalation suppresses the superconducting transition of the parent compound, NbS_2. From room temperature down to 2K the system remains metallic (figure). On the other hand, magnetic susceptibility measurement shows two mag-

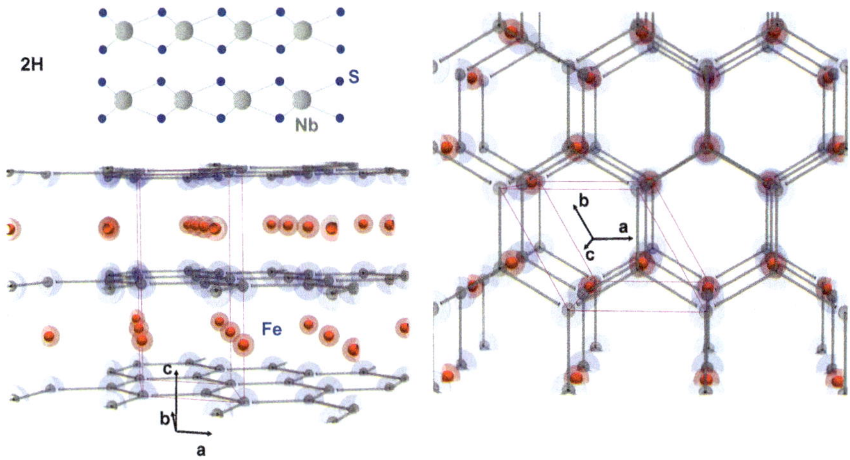

Fig. 4.1 **Structure of $Fe_{1/3}NbS_2$:** We see here two different view of the crystal structure where Fe ions are intercalated between NbS_2 layers which are 2H stacked

netic transitions: at $T_{N1} = 45K$ and $T_{N2} = 41K$, both of which host an antiferromagnetic state below the transition temperature. The in-plane resistivity and susceptibility (in-plane and out-of-plane) are shown in figure. Contrasting with the in-plane susceptibility measurement with the out-of-plane direction, it was expected to have the moments of the spins in the antiferromagnetic order aligned along the [001] direction.

The broken inversion symmetry of the unit cell makes it an ideal candidate for spin-orbit coupling, and subsequently for spintronic applications. To understand the role of inversion symmetry, let us consider an electron in a magnetic field $\mathbf{B}$ moving with a momentum $\mathbf{p}$. Then the force experienced by the electron is $\mathbf{F} = \frac{-e}{m}\mathbf{p} \times \mathbf{B}$. We can interpret this as an effective electric field experienced by the electron in its rest frame as

$$\mathbf{E}_{eff} = \mathbf{p} \times \mathbf{B}/m$$

By electromagnetic duality transformation, ($\mathbf{E} \rightarrow c\mathbf{B}$ and $c\mathbf{B} \rightarrow -\mathbf{E}$), then an electron in an electric field $\mathbf{E}$ will experience in its rest frame an effective magnetic field of strength:

$$\mathbf{B}_{eff} = \mathbf{E} \times \mathbf{p}/(mc^2)$$

The Zeeman splitting due to this magnetic field will include a term in the Hamiltonian describing such a system. Thus, in its rest frame where the electron experiences this magnetic field, the energy level will be split for two different spins of the electron with the symmetry allowed spin-orbit Hamiltonian $H_{SO} \propto \mathbf{E} \times \mathbf{p} \cdot \sigma$, where σ is the spin Pauli matrix. Since, this term in Hamiltonian must preserve time-reversal symmetry, under which $\sigma \rightarrow -\sigma$ and $\mathbf{p} \rightarrow -\mathbf{p}$, the term involving the momentum must also change its sign, i.e. $\mathbf{E} \times (-\mathbf{p}) = -\mathbf{E} \times \mathbf{p}$. Hence, in general, for $H_{SO} = f(\mathbf{E}, \mathbf{p}) \cdot \sigma$, $f(\mathbf{E}, \mathbf{p})$ must then be an odd-function of $\mathbf{p}$. This necessitates breaking the spatial inversion symmetry. Now that our set-up is ready to discuss the incredible spintronic phenomenology we dive straight into it.

Electrical Switching

The expected charge-spin coupling in this system is evidenced when the application of a direct current pulse switches the resistance of the material in its low-temperature antiferromagnetic ground state between two stable states. The experiment was conducted on an eight terminal device as shown schematically in the figure. The horizontal and vertical pathways were used for DC pulsing the crystal where an AC source of 100 μA fed current through one of the 45° channels. From the perpendicular channel to the AC channel voltage was measured to extract resistance. For a DC pulse of $5.4 \times 10^4 Acm^{-2}$ density for a duration of 10 ms across the blue electrodes, the authors observed a drop in resistance. A subsequent pulse across the red electrodes revived the resistance (Fig. 4.2).

Technologically, this was an important step towards antiferromagnetic spintronics and memory application. However, the phenomenon itself posed a scientific challenge in deciphering the mechanism of switching. As the system was warmed up above the Neel transition temperature, the switching amplitude gradually decreased to 0. Coupling to antiferromagnetic order is, thus, crucial for the phenomena. Understanding the antiferromagnetism in this material became of the highest importance. Out of many other possible mechanisms that involved the coexistence of spin glass phase with antiferromagnetic order, it was also suggested that the application of the DC pulse may serve to switch the sample between two distinct anti-ferromagnetic ground states that correspond to the two different resistance values in the measurement.

Antiferromagnetic Order

The out-of-plane susceptibility showed hints of the puzzle where it appeared to have two antiferromagnetic transitions. To gain a thorough understanding of the magnetic phase, neutron diffraction measurements along with structural and magnetic susceptibility measurements were carried out on single crystal samples of high quality. However, on top of the stoichiometric sample, this time the authors explored off-stoichiometric samples with iron site vacancies (x = 0.32) and interstitials (x=0.35). The stoichiometry was confirmed with energy-dispersive X-ray measurements.

Although magnetic defects are usually expected to suppress magnetic correlation and lower the transition temperature, highly tunable long-ranged magnetic phases were discovered as the Fe concentration was varied [2]. The wave vector of the under and over-intercalated samples differed from each other while the stoichiometric samples showed coexistence and competition of those phases. For the under-intercalated sample, the wave vector of the AFM order was found to be $\mathbf{k}_{stripe} = (0.5, 0, 0)$ and termed as the 'stripe' phase for the arrangement similarity of the Fe moments in

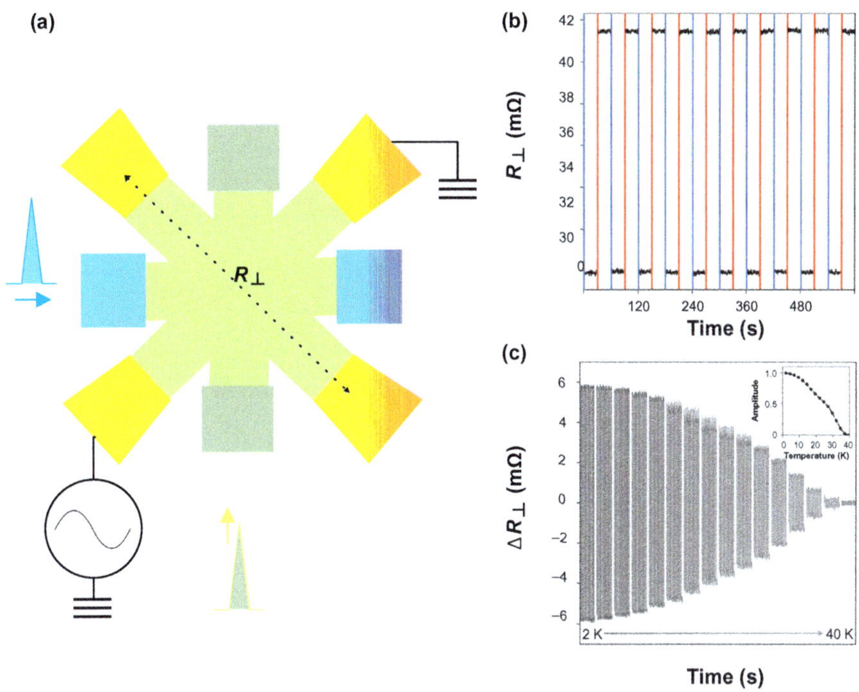

Fig. 4.2 **Electrical Switching of FNS coupled to the antiferromagnetic phase**: **a** The eight-pronged device used to measure $R_{\perp}$ by the authors [1]. They pulsed along the blue pads and the red pads. **b** shows the result of the DC pulses on the measured value of $R_{\perp}$. This resistance switching data is collected at 2K in the stoichiometric sample with x $=\frac{1}{3}$ **c** we have plotted the switching amplitude as a function of temperature (*inset* c) showing the collapse of switching amplitude above the Néel temperature of the sample

a layer with stripe pattern. For the over-intercalated sample, the wave vector was $\mathbf{k}_{sigzag} = (.25, .5, 0)$ and the AFM phase was named 'zigzag'. We will follow the same naming conventions in this thesis. The neutron diffraction measurements were carried out on individual single crystals with high crystalline quality. Fig shows the complex ground state phase diagram with the stripe and the zigzag phase dominating down to the lowest temperature near under-intercalated and over-intercalated samples, respectively. The magnetic moment arrangement in real space is depicted using a magnetic unit cell in three dimensions as well as shown in the triangular Fe lattice planes in between the TMDC layers (Fig. 4.3).

For the stoichiometric sample, they indeed observed both of these antiferromagnetic states. Around 45K first the stripe phase emerges and then around 41K the peaks related to the zigzag phase also show up. The emergence of the zigzag phase is met with a decrease in the intensity of the AFM peaks related to the stripe phase, indicating a phase competition for the real estate in the sample. This is consistent with the thermodynamic measurements, where the out-of-plane susceptibility measure-

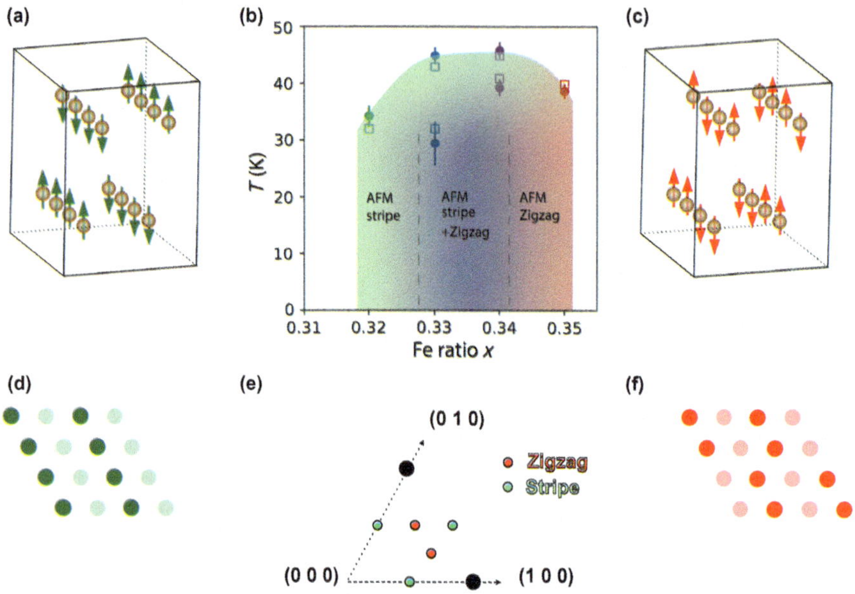

Fig. 4.3 **Phase Diagram of $Fe_x NbS_2$ from neutron scattering:** **a** and **d** shows the spin configuration in the 'stripe' phase on the Fe atoms from an isometric view and top view respectively. Similar diagram is shown for the 'zigzag' phase on the **c** and **f**. The dark(light) circles represent spin up(down). **b** panel shows the phase diagram as a function of temperature and Fe itercalation ratio, x adapted from [2]

ments saw indications of two antiferromagnetic transitions. We show here the key data from the publication where the temperature evolution of the magnetic peaks was measured. The peak positions for the magnetic peaks are shown in the figure along with their temperature evolution. The widths of the peak in all directions were instrument resolution-limited, affirming the three-dimensional structure of the magnetic domains. The authors also confirmed that the three-dimensional correlation remains preserved in the stripe phase even with phase competition with the zigzag phase in the $1/3$ intercalated samples. This means that the decrease in intensity observed in the stripe phase peak is not due to the correlation changing from a 3D to a 2D peak shape (Fig. 4.4).

Furthermore, all of the phase transitions were continuous phase transitions. Extracted critical exponent β from the temperature evolution of the magnetic peak intensity showed consistency with 2D Ising model. In the following we tabulate their findings about the antiferromagnetic phase in each stoichiometric sample in table 4.1. We also tabulate their lattice parameters for competition. We see that the effect of intercalation only changes the out-of-plane lattice parameter.

The neutron scattering also revealed that the Fe-spin moment orientation is along the c-axis for all samples with negligible in-plane components. This ruled out the spin-glass mediated switching mechanism proposed, since, an in-plane directionality

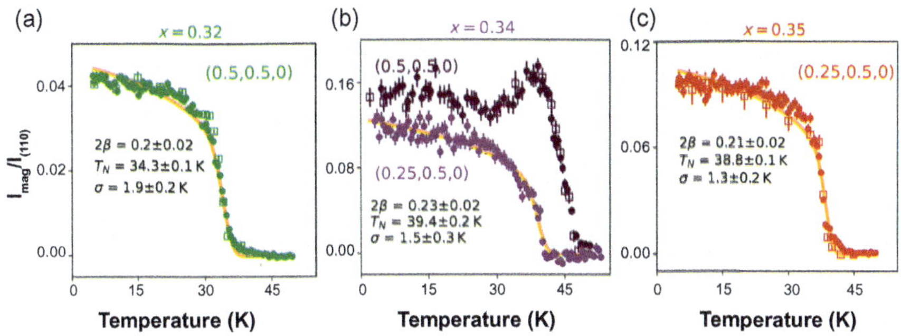

Fig. 4.4 Neutron Scattering result [2]: This shows fundamentally the interesting phase competition in the stoichiometric sample from neutron scattering. **a** The under-intercalated sample (x = 0.32) shows only the stripe phase **c** the over-intercalated samples (x = 0.35) shows the 'zigzag' phase. Both of which follows the mean field description of an Ising model with the critical exponent and Néel (T_N) mentioned. σ is the spread of T_N. For the stoichiometric sample (x $\approx$ 1/3) we see the 'stripe' phase turns on at a higher temperature first followed by emergence of 'zigzag' phase at lower temperatures where phase competition ensues

Table 4.1 Key results from the neutron scattering study [2]

Attributes	$x < \frac{1}{3}$	$x \approx \frac{1}{3}$	$x > \frac{1}{3}$
T_N (K)	34	$\sim 45(1^{st})$ $\sim$41(2nd)	39
β	0.2	0.23	0.21
k_{AFM}	(0.5, 0, 0)	(0.5, 0, 0)-1st (0.25, 0.5, 0) - 2nd	(0.25, 0.5, 0)
a = b (Å)	5.76	5.76	5.76
c (Å)	12.13	12.15	12.19
Space group	$P6_322$	$P6_322$	$P6_322$

of the spin-orbit torque would not be able to switch one domain to the other, both having moments along out-of-plane direction.

In summary, this sample hosts a strong charge carrier to antiferromagnetic order coupling, where the antiferromagnetic phase itself has two possible ground states: stripe and zigzag. They dominate the low-temperature ground state for under-intercalated and over-intercalated samples, respectively. However, for the stoichiometric sample, they compete and coexist. Thus, in these samples, an external DC pulse may switch the configuration in favor of one phase switching the resistance.

This brings us to the central question we will try to answer using coherent resonant soft X-ray scattering:

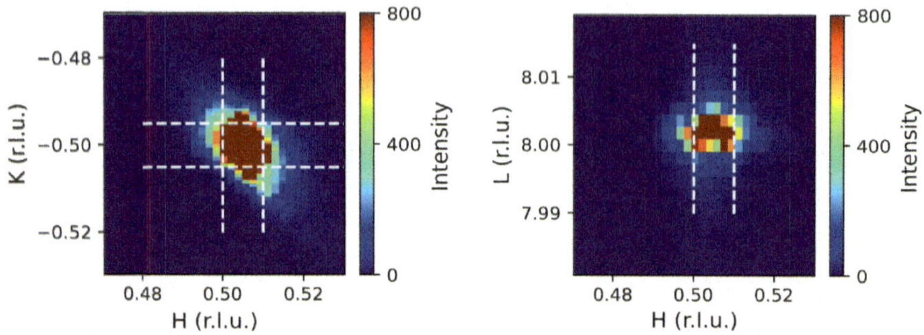

Fig. 4.5 Observed Charge Order peak in the detector: The intensity of the superlattice peak sampled on the HK plane (left panel), and on the HL plane (right panel)

How do the two low-temperature phases compete with each other where the two phases break the high-temperature symmetry differently in a strongly correlated environment?

Before we start the discussion, let us, first, visit the concomitant charge ordering phenomenon associated with the spin ordering in these samples to fully unveil the complexity of these materials.

4.3 Discovery of Charge Order and Coupling to AFM Phase

With the strong spin-charge coupling observed in this material and suppression of superconductivity with intercalation, we asked the obvious question of whether the antiferromagnetic order could stabilize any charge order. The parent compound is peculiar in the family of TMDCs by not exhibiting any charge density wave order. It was suggested that the anharmonic fluctuation of the lattice destabilizes the charge order in the parent compound , which was otherwise close to an instability to charge order due to many-body effects involving competing Coulomb and electron-phonon interaction . Furthermore, a recent experimental investigation of diffuse scattering of hard X-rays showed faint charge order peaks at M points i.e. (0.5, 0, 0) wave vector positions attributed to either Friedel oscillations around impurity or extremely faint charge density wave order . We were further lured to search for charge order in intercalated compounds due to the small resistivity kink observed at the Neel transition temperature, and optical polarimetry measurement showed three-state nematicity related to the measurement of optical birefringence.

Hard X-ray Scattering experiments were carried out using the psi-circle diffractometer of the 6-ID-B beamline at the Advanced Photon Source, Argonne National Laboratory with an incident photon energy of 10 keV and σ-polarization. The scattered beam was detected with a Dectris Pilatus 100K area detector without polariza-

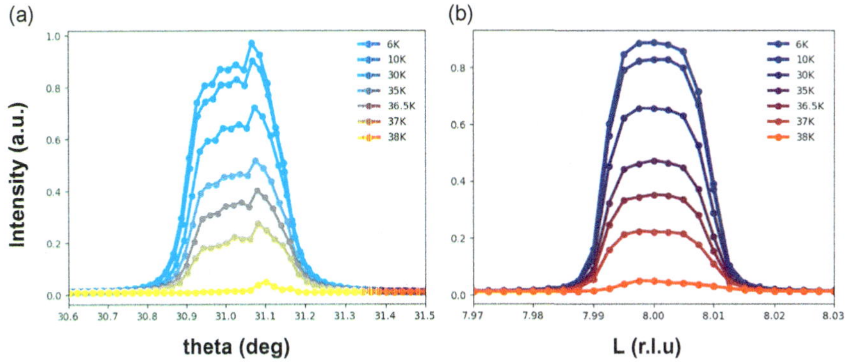

Fig. 4.6 Temperature evolution of the emergent charge order: a rocking curve scans on the (0.5, −0.5, 8) peak **b** L-scans to understand the out-of-plane correlation length on the same peak

tion analysis. We used high-quality single crystals with Fe ratio of $x = 0.32$ and 0.35, previously measured with neutron scattering . Their mosaicity was determined to be $0.15°$ in peak width (full width at half maximum, FWHM) on the $(1, 0, 7)$ atomic Bragg peak. For the $x = 0.32$ sample, we also examined a second small crystal with better mosaicity, with FWHM $= 0.02°$. The samples were glued to a Cu-post using silver paint and mounted one at a time on the closed-cycle cryostat with Be domes for the X-rays to pass through. The temperature was changed between 5.6 K and 50 K with an instrument resolution and control of 0.1 K.

We first studied the X-ray scattering on the $x = 0.35$ sample, which with neutron diffraction showed a zigzag order with a magnetic wave vector $k_{m2} = (0.25, 0.5, 0)$ below $T_N = 38$ K. At $T = 6$ K, we observed new superlattice peaks at $Q = (0.5, 0, \mathrm{L})$ and (0.5, 0.5, L) (L = integer values) as seen from the detector image in Fig. 4.5. We measured six equivalent momentum positions for each Q and confirmed a series of superlattice peaks with half-integer in-plane indices and integer L. At $T = 45$ K, all those peaks disappear, revealing a temperature-dependent behavior. We recognize these super-lattice peaks as the charge order peaks. In what follows, we characterize the intensity to justify our recognition (Fig. 4.7).

Firstly, we note that the peak showcases a three-dimensional correlation. In Fig. 4.6 we demonstrate the melting of this super-lattice peak from rocking curve scans and L-scans at the peak position of (0.5, −0.5, 8) taken at different temperatures as we warmed the sample from base temperature. A well-defined FWHM along the L direction provides further evidence of the three-dimensionality of the charge order.

To characterize the three-dimensional correlation length of the super-lattice order, we have taken reciprocal cuts of a representative peak (0.5, −0.5, 8) equivalent to the (0.5, 0, L) peaks along different high-symmetry directions as shown in Fig. 4.8. To evaluate the domain size from these cuts, we have taken the values of the ends of the FWHM positions on the peak and evaluated $\Delta q = |(H, K, L)_{end2}| -$

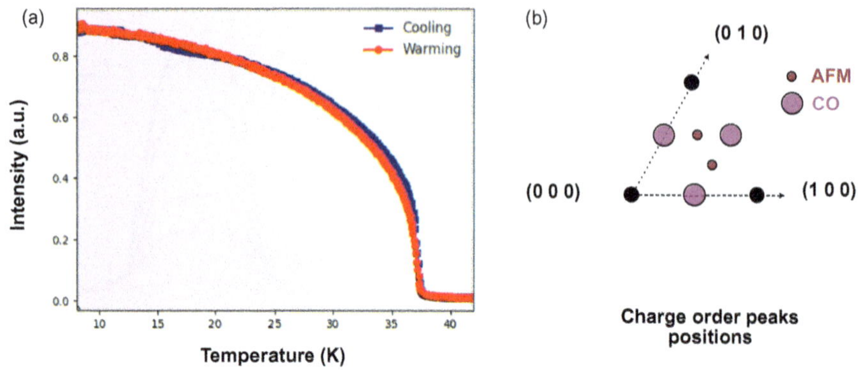

Fig. 4.7 Temperature evolution and Reciprocal location: a warming and cooling on the charge order peak showing no hysteresis and the phase transition to be continuous in nature. The onset temperature is exactly same as the Neel temperature of the sample indicating an antiferromagnetic origin of the charge order **b** position of the charge order peaks and the AFM peaks below the transition temperature along with the Bragg peaks for over-intercalated samples

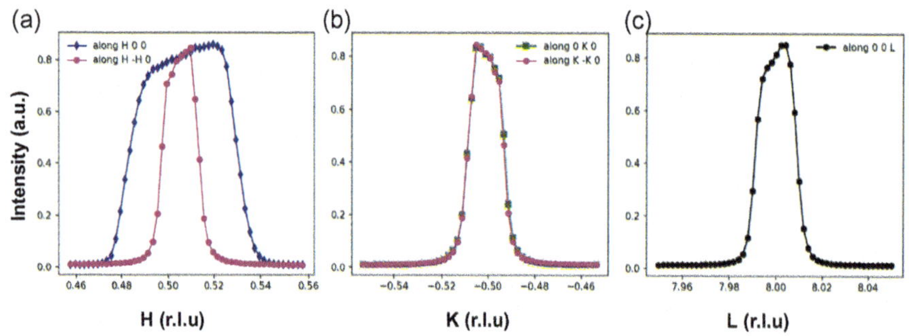

Fig. 4.8 Correlation length along different high symmetry direction of the emergent charge order: different cuts along the (0.5, −0.5, 8) peak is shown in the panels and used to extract the correlation length along the directions H (**a**), K (**b**), and L (**c**). The correlation is three dimensional but highly ansiotropic

$|(H, K, L)_{end1}|$. Then we converted it into real space measurements using $2\pi/\Delta q$. The extracted values are tabulated below in Table 4.2 and shown in Fig. 4.9.

We observe that the in-plane correlation lengths are longer by a factor compared to the out-of-plane correlation length. It is not surprising that the out-of-plane correlation length is smaller. Furthermore, these peaks with half-integer in-plane indices are visible at both even and odd values of L, contrary to the selection rule of odd L values for the zigzag and stripe magnetic phases, where two adjacent Fe layers stack anti-parallel. Thus, the superlattice order is equally present in every layer of the unit cell. What is more surprising is that the in-plane correlation lengths are also not identical along the high-symmetry directions. As shown in the table and visualized

Table 4.2 Correlation length of the charge order evaluated from Fig. 4.8

	Δq (Å^{-1})	Real space size (nm)
Along (100)	0.0039	162.3
Along (010)	0.0014	436.7
Along (1-10)	0.0028	224.2
Along (001)	0.0084	74.8

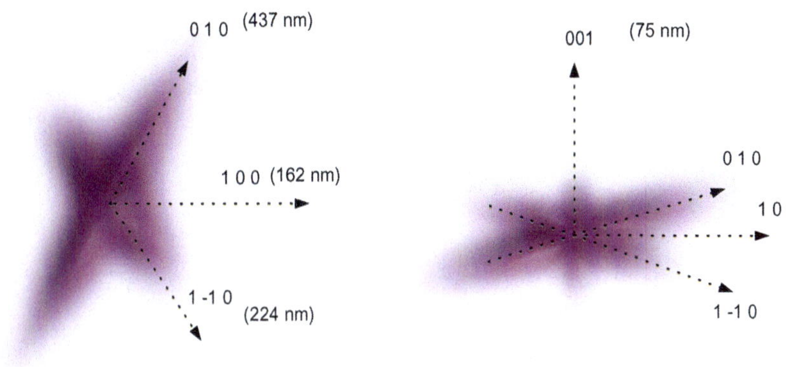

Fig. 4.9 Anisotropic correlation length of the charge order: (left) top view (right) isometric view of domain shapes following the correlation lengths evaluated. The lengths are marked along the crystallographic axes

in the Fig. 4.9. There seems to be a preferential direction for a longer correlation length.

We contrast the temperature evolution of this superlattice peak with the intensity of a Bragg peak of equivalent d-spacing. After alignment to the Bragg peak we collected rocking curve scans on the peak as we changed the temperature and observed no significant change in the peak as shown in the Fig. 4.10. We have also checked the intensity at the antiferromagnetic peak location to no avail. We scanned over other high-symmetry positions with wave vectors $(n/3, m/3, L)$, $(n/3, 0, L)$, $(n/4, m/4, L)$, and$(n/4, 0, L)$ with $n, m \in \mathbb{Z}$ and $L = 7, 8$ at the lowest temperature with the same polarization of the incoming beam, but we did not find any other super-lattice peak.

Secondly, to investigate the electronic origin of the super-lattice peak, we moved to the energy space. We performed fixed-q energy scans on the peak at Fe-K edge and Nb-K edge with incoming polarization set to σ. In addition, we analyzed the scattering signal to disentangle the charge component of scattering from the spin. These experiments were carried out at the 4-ID beamline of National Synchrotron Light Source-II of the Brookhaven National Lab. We used a six-circle diffractometer in four-circle mode with fixed phi and a point detector to collect the scattered signal.

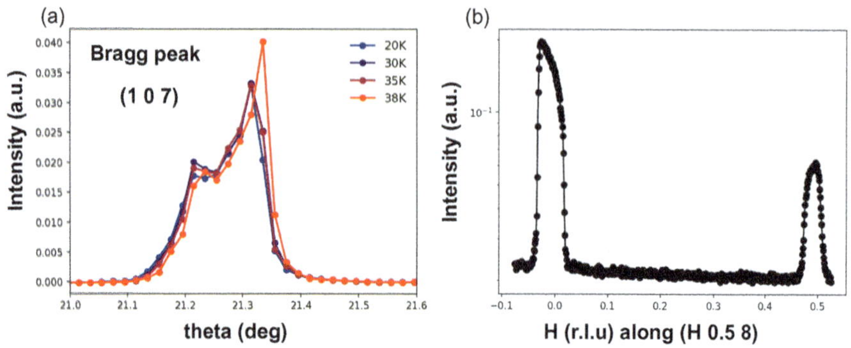

Fig. 4.10 Other relevant Bragg peaks: a Temperature evolution of the Bragg peak (1 0 7) showing no change with temperature **b** Null intensity at 'stripe' order peak position (0.25, 0.5, 8) at 6K

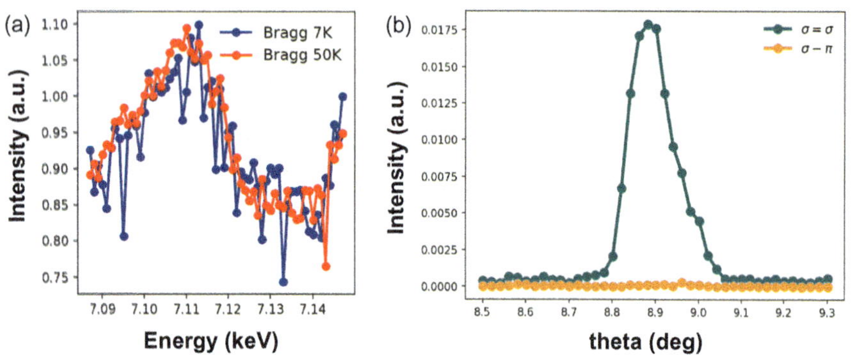

Fig. 4.11 Energy dependence of the Bragg peak and Polarization dependence of the charge order peak: a Energy dependence from fixed-q energy scan on the Bragg peak shows no temperature dependence. It shows a decrease in intensity detected from scattering as it hits the Fe-K edge decreasing the penetration depth due to increased absorption **b** the charge order peak clearly sensitive to the $\sigma - \sigma$ channel proving a charge origin to the superlattice peak

We normalized the energy-integrated scattered intensity but resolved the polarization of the scattered beam using an analyzer crystal of Al-660 for Nb-K edge and Cu-220 for Fe-K edge.

We note that almost all of the scattered signal is observed in the $\sigma - \sigma$ channel, i.e., where the incoming and outgoing beams are σ polarized. This channel is dominated by the charge-scattering component. A change in polarization will indicate a spin contribution to the scattered signal. The absence of the peak intensity in the $\sigma - \pi$ channel thus excludes the magnetic origin of the super-lattice peaks.

In the same Fig. 4.11, we have also shown the fixed-q energy scan on a Bragg peak around the Fe-K edge of 7.12 keV. During fixed-q energy scans, the energy was varied using a double crystal monochromator of Si(111). The energy resolution is 1 eV at 10 keV. At the resonance, the intensity only decreases reflecting absorption

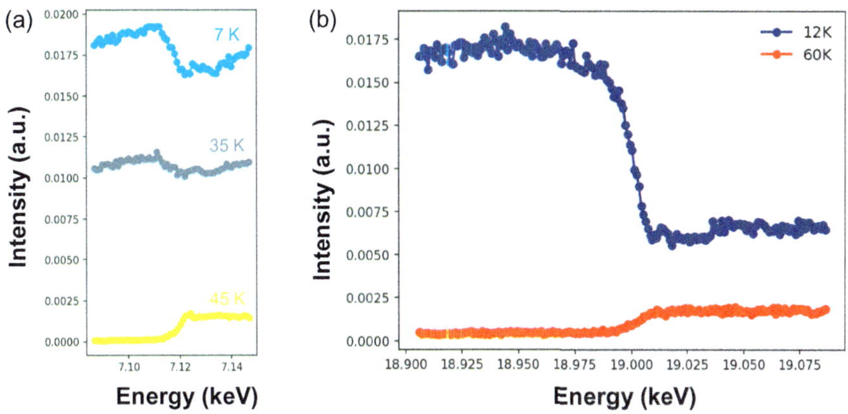

Fig. 4.12 Energy dependence of the charge order peak near Fe-K edge and Nb-K edge as we changed the temperature: a Fe-K edge and **b** Nb-K edge energy dependence with incoming polarization selected as σ and the temperature was varied. The measurement was done with fix-q energy scan on the peak location. The high temperature measurements do not host a Bragg peak and thus mimic the fluorescence data

by the layers close to the surface which decreases the scattering volume, and hence the intensity. This is a usual response at resonance energy conditions if there is no electronic origin of the peak, such as the Bragg peak that we investigate here. The spectra do not change below the transition temperature as well. We observe similar responses at the charge order peak as discussed below denoting a lack of electronic origin such as valence or orbital ordering (Fig. 4.12).

The low-temperature energy responses across Fe-K edge at 7.12 keV or Nb-K edge at 19keV, are identical with the energy response of the Bragg peak. Above the transition temperature, there are no peaks. However, when we scanned at the peak location the intensity now increased across the Fe-K edge or Nb-K edge due to an increase in the fluorescence yield. In both of these cases, we created a core hole at the 1s state and let it collapse. we observed the photon emitted in the process of the core-hole collapse. A lack of energy modulation near the Fermi level by the valence environment (like we saw in the case of nickelates) may mean that the electrons in the cations do not participate in the charge ordering.

It is worth asking the question: why did we not observe the charge order peak in neutron scattering investigation? Where the relative intensity of the magnetic peaks was about 50% of the Bragg peak intensity, the half-integer peaks on the other hand were found to be approximately 3 orders of magnitude smaller than the Bragg peak intensity. Thus, these peaks may have produced a negligible relative intensity above the background level in neutron scattering. That is probably one of the reasons for not seeing the peaks in neutron scattering. A strong charge ordering influences structural distortion, resulting in a shift of the whole atom, like a bond distortion in nickelates. We have reasons to believe that the valence electrons of the cations do not participate in the process. Then the cations are more likely to scatter the neutrons strongly but

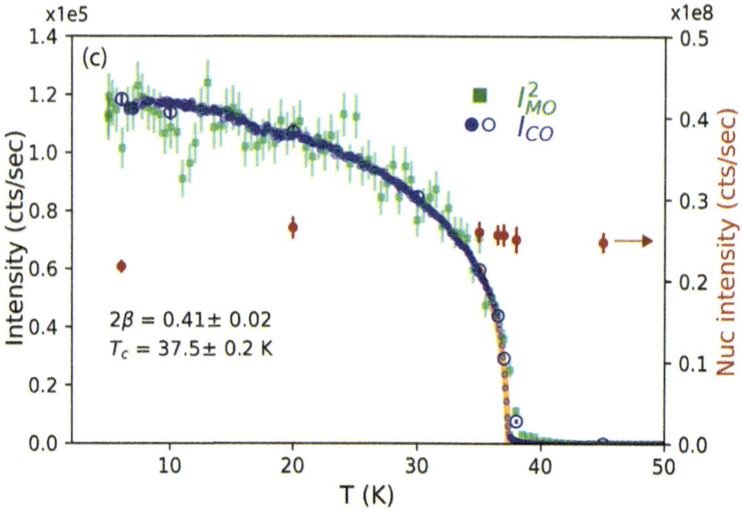

Fig. 4.13 Correspondence between magnetic and charge order intensity: we show here the charge order and square of the magnetic order parameter with varying temperature. They correlate exactly indicating a direct proportionality. Contrasting with these temperature evolution the structural Bragg peak does not change with temperature

do not undergo the distortion required and as a result, this peak does not show up. In hard X-ray scattering, we scatter from the electrons and the electronic modulation shows up as a Bragg peak.

In contrast, we did not observe any additional temperature-dependent superlattice peaks in the samples with x=0.32 that host the magnetic 'stripe' phase. We scanned over roughly 36 peaks with half-integer in-plane indices, as well as other high-symmetry directions, including $(n/4, m/4, L)$, $(n/3, m/3, L)$, $(n/4, 0, L)$ and $(n/3, 0, L)$ (n, m = integers, L = 6,7). We studied two $x = 0.32$ samples: one neutron sample with a mosaicity comparable to the $x = 0.35$ sample (FWHM = 0.15°), and the second one with a sharper peak width (FWHM = 0.02°). Both samples did not reveal any apparent temperature-dependent peak features. The superlattice peak at (0.5,0.5,L) persists up to room temperature (see Fig. S1 (b)), most likely due to a small portion of the Fe occupancy at 2b sites in the under-intercalated samples. Given the surveyed momentum positions, the invisibility of charge ordering in $x =$ 0.32 is plausibly related to highly tunable magnetic structures with respect to the Fe ratio x, which will be discussed later (Fig. 4.13).

The Magnetic Link

A surprising connection of charge order to magnetism was found from the temperature dependence of the magnetic order. Our argument follows the temperature dependence of the charge order peak in the sample with x = 0.35. We determined the transition temperature for the charge order phase by fitting the order-parameter curve with a power law exponent in the temperature range above 30 K. The order parameter plot was collected by warming or cooling the sample at a constant rate while recording the charge peak intensity in, the so-called timescan. The data is shown in figure . The fit yielded a charge order transition temperature of $T_C = 37.5(2)$ K and a power law exponent for the charge ordering of $2\beta = 0.41(2)$. From the magnetoelastic model to be discussed in the following, this value implies a 2β value of 0.21 for magnetic ordering, consistent with the 2D Ising model value of $2\beta = 0.25$. The value of T_C is in good agreement with the magnetic order transition temperature, suggesting a strong connection between charge and magnetic order.

The charge order in the zigzag phase is clearly driven by the antiferromagnetic order. To further understand the intimate relationship between the charge and magnetic orders, we compared the temperature-dependent intensities of the two orders. Interestingly, the square of the magnetic intensity I_{MO}^2 matches perfectly with the charge order intensity I_{CO} after scaling by a constant factor. This correlation can be attributed to the charge modulation from magnetoelastic coupling between the lattice and magnetism [3], which induces uniform strain ε via the coupling term:

$$\sum \varepsilon_{i,j} m_i m_j. \tag{4.1}$$

The free energy for the magnetic order is in the form of

$$\mathcal{F}(m) = -a_m(1 - \frac{T}{T_N})m^2 + b_m m^4, (a_m, b_m > 0) \tag{4.2}$$

. Due to the magnetoelastic coupling, there is an additional term $\Delta\mathcal{F}$ in the free energy [4],

$$\Delta\mathcal{F} \sim -\gamma \rho m^2 + \frac{1}{2}\rho^2, \tag{4.3}$$

where m is the magnetic moment, γ is the coupling strength and ρ is the modulated charge density, which is a secondary order parameter. This coupling leads to the charge scattering intensity from minimizing the free energy (Fig. 4.14b) as:

$$I_{CO} \sim \rho^2 \sim m^4 \sim I_{MO}^2. \tag{4.4}$$

This agrees with the empirical intensity relationship in our data.

In Q-space, the free energy can be expressed as:

$$\sum_{k_1,k_2} \rho(-k_1 - k_2) m(k_1) m(k_2) + \frac{1}{2}\rho^2 \tag{4.5}$$

This formula implies that the wave vector of the charge order, k_{CO}, would be the sum of two magnetic wave vectors, k_m. As shown in Fig. 4.14c in the $x = 0.35$ sample, 12

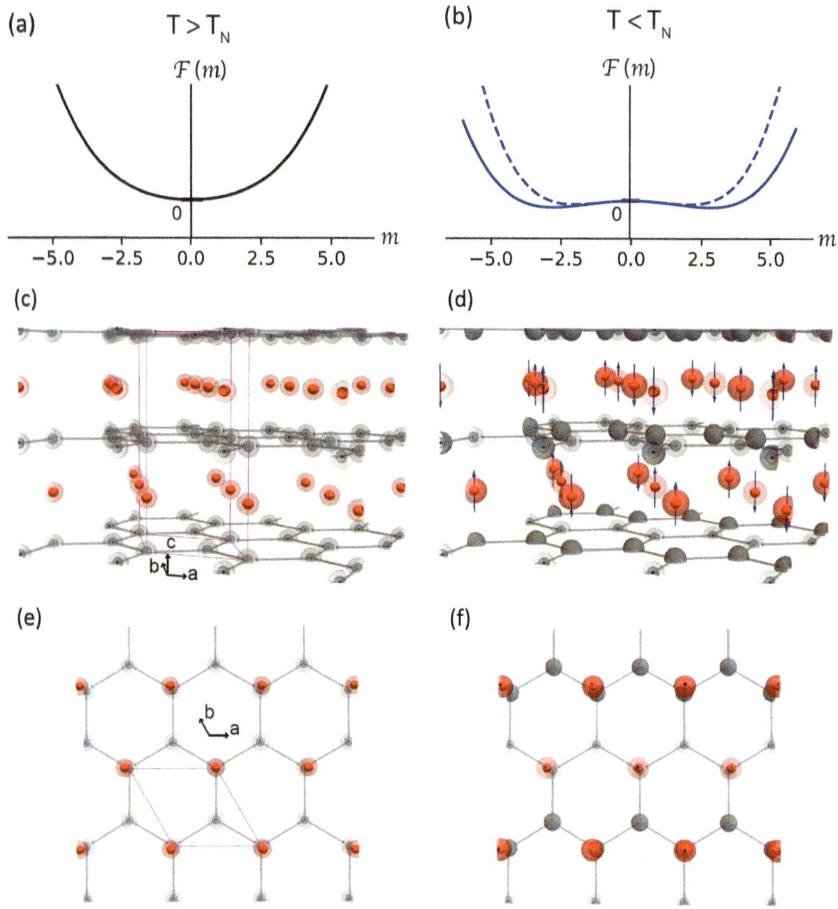

Fig. 4.14 Free energy and electron density modulation at low temperatures: Schematic illustrations of the free energy for **a** $T > T_N$ and **b** $T < T_N$. In panel **b**, the dashed blue line is the free energy for the antiferromagnetic order considering it undergoes a continuous phase transition. The solid blue line includes the change of the free energy $\Delta\mathcal{F}$ due to magneto-elastic coupling, lowering the ground state energy. **c** and **e** High temperature ($T > T_N$) lattice showing only Fe (red) and Nb (dark) atoms in three-dimensional and ab-plane projections, respectively. The pink line denotes the unit cell. **d** and **f** Illustration of the 3D charge ordering ($T < T_N$), accompanied by the zigzag spin ordering in three-dimensional and ab-plane projections, respectively The dark and light red (grey) spherical clouds denote the charge density modulation localized on Fe and Nb sites, and blue arrows mark the spin structure of Fe

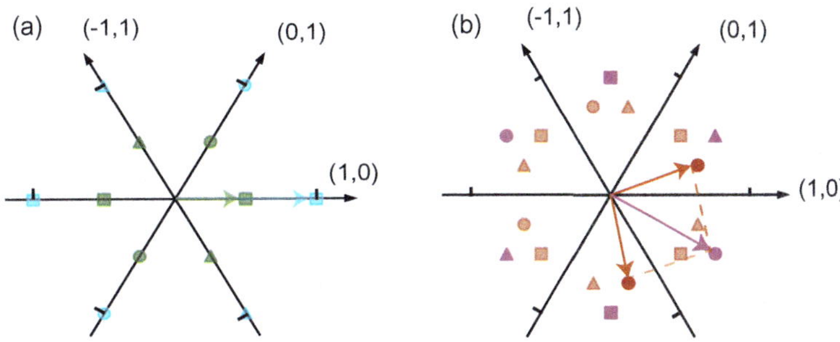

Fig. 4.15 Charge Order and Magnetic order relationship in reciprocal space: different shapes of the peaks corresponds to different domains **a** the charge order and magentic order peaks for the under-intercalated samples. The relationship places the charge order peaks exactly on top of the Bragg peaks **b** the charge order peak for a 'zigzag' phase can be found by adding the wave vectors of two magentic peaks

equivalent magnetic wave vectors are assigned to three magnetic domains (denoted by three symbols). Within one domain, two k_{m2} (red vector) can generate one charge wave vector k_{CO2} (blue vector), leading to a total of six equivalent wave vectors associated with (0.5, 0.5, 0). This naturally explains the observed peak positions with half-integer in-plane indices considering the momentum positions $Q_{CO2} = \tau \pm k_{CO2}$ (τ is the atomic Bragg position).

For the stripe phase, the magnetic wave vector occurs at half-integer values (Fig. 4.15 (right)). Any charge scattering originating from the magneto-elastic coupling would be orders of magnitude weaker than the regular Bragg scattering from the crystal lattice. Doubling of k_{m1} causes the charge order peaks to appear at the atomic Bragg peak positions, making them undetectable in the experiment. Consequently, we are unable to determine if any magnetoelastically induced charge ordering in the $x = 0.32$ stripe phase sample.

The magnetic order induced charge ordering in $Fe_{0.35}NbS_2$ is different from that of other pristine TMDs exhibiting incommensurate charge-density-wave order, such as 2H-$NbSe_2$. 2H-$NbSe_2$ was reported to display the low-energy Fermi surface nesting [5–11]; but later the CDW formation was found to be attributed to the electron-phonon coupling mechanism [12, 13]. In $Fe_{0.35}NbS_2$, our ARPES measurements did not reveal any clear evidence of CDW features via auto-correlated Fermi surface analysis [14] (see Fig. S3). In addition, our spectra have revealed a strong photon energy dependence of the electronic structure caused by the Fe intercalation, distinct from the quasi-2D electronic band in bulk NbS_2 [15]. These results confirm that the charge order, here, is not a simple low-energy electronically-driven transition, but rather the intra/-inter-layer magnetic exchange coupling involving Fe is a vital factor.

Magnetic Field Dependence of the Charge Order

An obvious consequence of the coupling of the charge order and the magnetic order parameter is that the charge order parameter should respond to the external magnetic field. To experimentally verify this consequence we conducted magnetic field dependence hard X-ray scattering. We conducted these experiments at the P09 beamline of Petra-III light source in DESY-Hamburg. Field-dependent X-ray scattering constrains the accessible Ewald sphere due to the experimental geometry and access to the scattered beam through the magnet. At this beamline, we have a 14T superconducting magnet which provides the field in the vertical direction in a horizontal scattering plane. The scattered intensity was recorded using an area detector. To set up for the experiment, we had to build a sample holder and mount the samples beforehand to meet with the dimensions of our samples. We mounted the 0.35 intercalated sample which we investigated in neutron and hard X-ray scattering. We showed in Fig. 4.16 our mounting schema and the sample holder with multiple sample assemblies. The diffractometer was essentially a two-circle diffractometer with no χ and Φ motor. Investigation on other samples was time-restricted so we present here the data from the 0.35 sample only.

At the Bragg condition, we accessed the (0.5, 0, 8) peak at the base temperature. The end station offered a base temperature of 3K. Incoming polarization was set to σ, and energy was set to 10keV. Firstly from the temperature evolution, we observed a previously unforeseen decrease in intensity at the lowest temperature. Maximum intensity was observed around 15–20 K. Secondly, the correlation length also changes

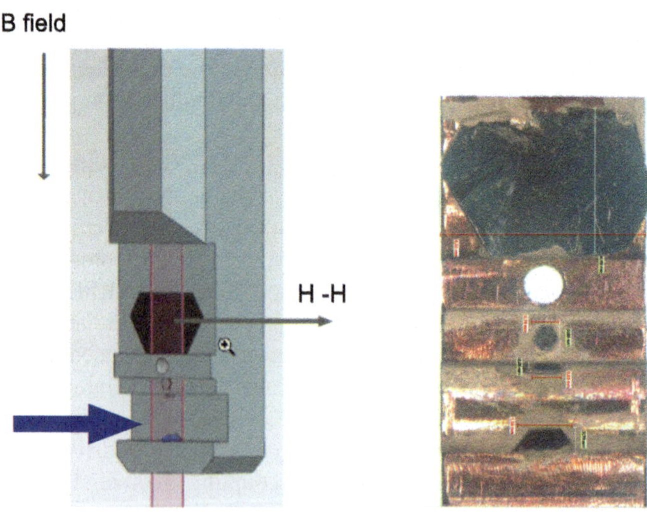

Fig. 4.16 Sample holder for field dependence of the charge order: (left) schematic of the scattering set-up with the external field pointing along the in-plane direction. (right) Actual samples mounted before putting into the diffractometer

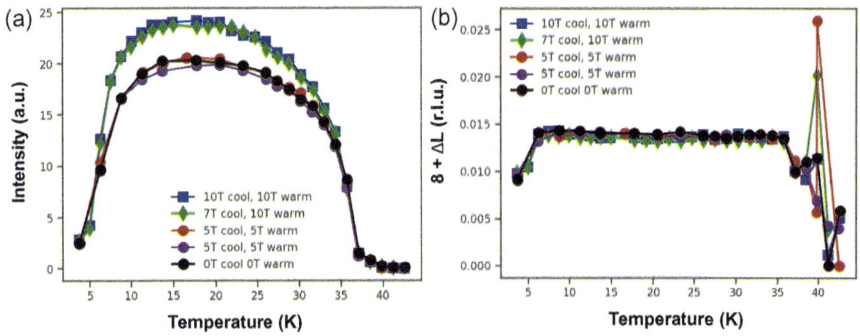

Fig. 4.17 Field Cooling of the Charge Order: We have shown here multiple temperature cycle with different external field as mentioned. With external field the transition temperature does not change indicating the short range correlation remains preserved. On the right panel we see the correlation length in the out-of-plane direction for different field cooling showing identical correlation length for all

by 34% indicating a melting of the charge order. We need further measurements to test the reproducibility of these low-temperature phenomena.

We followed this measurement up with field-cooled charge order formation and then observed the temperature dependence of the charge order while keeping the field on. Figure 4.17 shows various such field-cooled and field-warmed measurements. For each temperature, we collected rocking curve scans and L scans on the peak. The intensity plotted here is the area under the curve of the beam-current normalized signal for the L scans. The rocking curve measurements show identical development. They follow the same intensity evolution until 5T, however, we observe an increase in intensity without any change in out-of-plane correlation length denoting the amplitude of the order parameter changes as we expected from the coupling term.

Then we wiped the remnant of field dependence on the order parameter by warming and cooling without an external field. Finally, we set the temperature at 20K where the order parameter was maximum during the temperature cycle, ramped the field slowly to a desired field, and held it fixed when we collected data. We collected rocking curve scans at every data point shown in the Fig. 4.18. Finally in Fig. 4.18 we plotted the area under the curve, which also showed field-dependent intensity change.

Finally, we have unfolded the complete complexity of this material. There are two magnetic phases 'stripe' and 'zigzag' that compete with each other at the stoichiometric sample. Furthermore, the charge order phase couples with the magnetic order phase. The magneto-elastic coupling term fixes the charge order peak position in the reciprocal space. For the 'zigzag' phase we find half-order peaks whereas the 'stripe' phase only allows charge-order peaks to coincide with the Bragg peak positions.

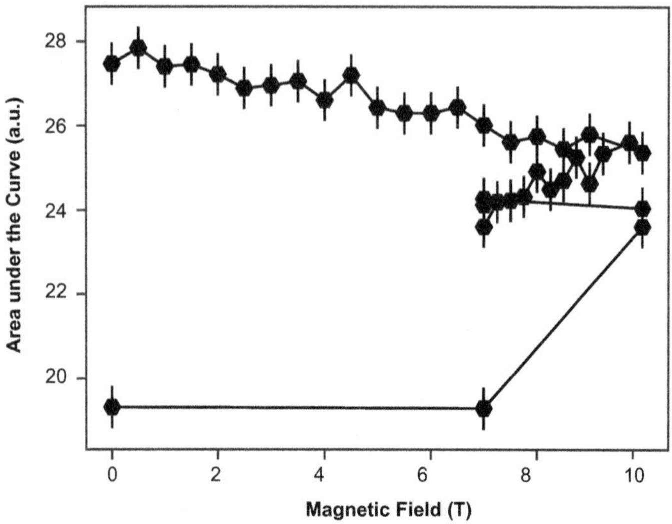

Fig. 4.18 Field dependence of the charge order at 20K: Increasing the field increases the area under the curve of the charge order. The two different points at 10T is found when we increased the field upto 12T after the lower intensity point. Field is pointed perpendicular to the scattering plane

4.4 Phase Competition from Coherent Scattering

To explore the phase competition between two anti-ferromagnetic orders we carried out coherent resonant soft X-ray scattering on a sample with x=0.327. We determined the intercalation ratio from an average of 10 measurements of energy-dispersive X-ray measurements. Since resonant coherent scattering is sensitive to the phase texture of electronic order, under competition a change in phase texture will be reflected on the speckle pattern to be analyzed to characterize the competition. The experiments were carried out in the beamline of 23ID CSX at the National Synchrotron Light Source-II of Brookhaven National Lab. The scattering chamber is identical to the one discussed in Chap. 3. To do the resonant scattering we tuned our incoming beam to the resonance of Fe-L3 edge. However, at this low energy, the accessible reciprocal space is very small. Even though we could access the (0.5, 0, 0) peak i.e. the stripe phase antiferromagnetic peak, the 'zigzag' peak at $q = (0.25, 0.5, 0)$ peak stayed out of range. Hence, for the soft X-ray resonant scattering experiment we will focus only on the 'stripe' phase antiferromagnetic order. We will follow the evolution of its phase texture to understand the melting of this phase (Fig. 4.19).

To be certain of the position we are measuring, we had to adapt some sample design. Sometimes, there are sample features, such as defects on the surface, which act as a position marker. However, it is always a better idea to create one, which will not change as we change the temperature. To create the marker we used gold wires of 25 μm gauge. First, we fixed the sample on a gold-plated Si substrate and then

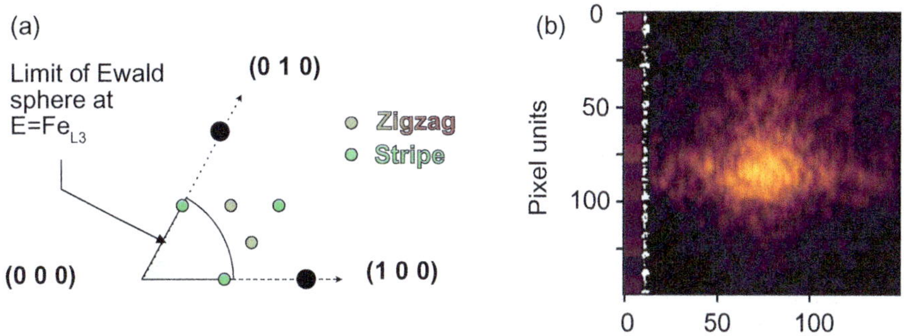

Fig. 4.19 Restricted Ewald sphere at Fe-L3 edge and speckle pattern a We show both 'zigzag' and 'stripe' phase peak locations with the shaded section indicating the Ewald sphere at Fe-L3 edge. We can access some 'stripe' phase peaks but can not access any 'zigzag' phase peak. **b** the speckle pattern on the detector for the Stripe phase at 41.2 K

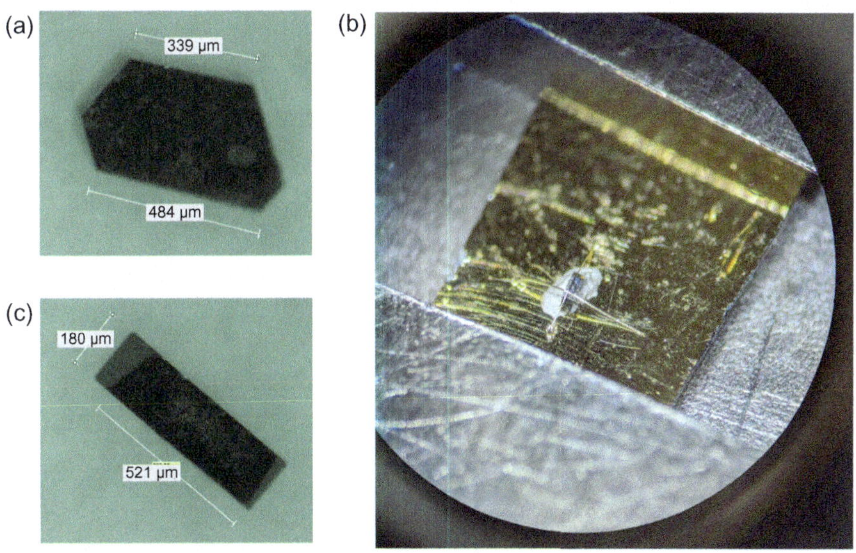

Fig. 4.20 Sample mounting preparation: a, c show the dimension of the sample we investigated in the coherent soft X-ray scattering experiment. **c** shows the gold wire tied over the sample which acted as fiducial marker

tied the gold wires to the plating after tightly winding over the sample. We wound two straight wires to make it look like axes as shown in the Fig. 4.20. We used the crossing of the wires as the origin of the sample positioning.

4.4.1 Melting of the Stripe Phase

The 'stripe' phase melts in two ways: firstly, as we warm up, it melts due to thermal fluctuation, and secondly, as we cool down, it melts due to competition with 'zigzag' phase. From the susceptibility measurements, we know that the stripe phase sets in at $T_{stripe} = 45$ K, whereas the 'zigzag' phase sets in at $T_{zigzag} = 41$ K. We set the liquid helium flow to reach the base temperature of 20 K and then used a heater to go to the desired sample temperature. The temperature control was accurate down to ± 5 K.

The first question we ask: *Are the two melting similar?*

To illustrate this point, we show here a schematic of an antiferromagnetic domain that first forms as we lower the temperature. For a continuous phase transition, the fluctuations are long-range correlated at the transition temperature. This is evidenced by the susceptibility divergence at T_C. As we lower the temperature this correlation length decreases to a finite value and the ordered patch expands. Thus, under thermal fluctuations, we expect the ordered patch to melt from the boundary. For a complex order, parameter field, even though the amplitude fluctuations stabilize below T_C the phase factors remain free to fluctuate (Fig. 4.21).

On the other hand, melting due to phase competition can happen either from the boundary or the core of the patch. For a melting from the boundary, the 'zigzag' phase will nucleate at the domain walls of the 'stripe' phase. It will look like thermal melting of the 'stripe' phase. Recently the authors of [16] have shown that domain walls of a quasi-2D charge density wave order with wave vector (0.5, 0, 0) can nucleate a superconducting phase known to compete with the charge density wave phase. Unless there is a specific way the new phase nucleates and grows, this melting will appear identical to thermal melting in terms of the phase texture of the initial phase.

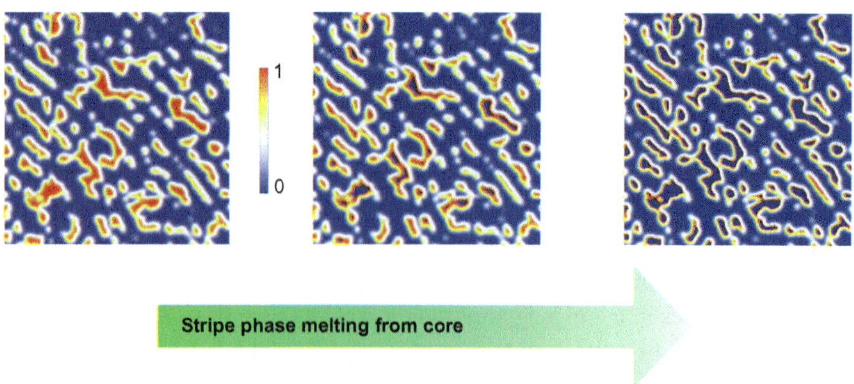

Fig. 4.21 Schematic of domains melting from boundary: As thermal fluctuations increase, we argue that the order starts to get washed away where the order parameter strength will be the lowest, i.e. at the boundary

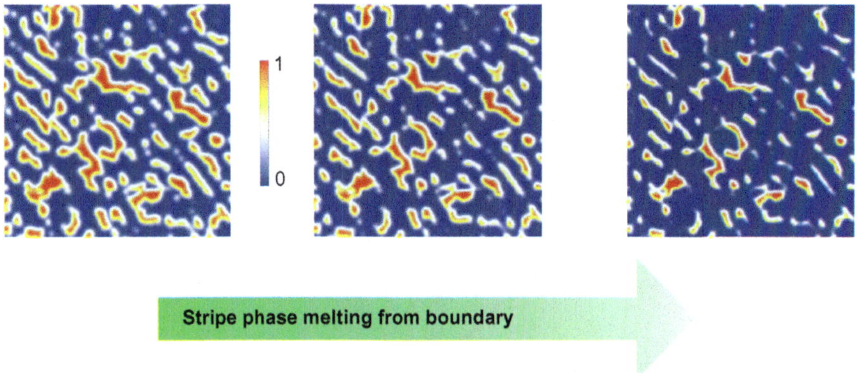

Fig. 4.22 Schematic of domains melting from core: Another way an order can melt if the new competing phase nucleate where the existing order parameter is the strongest, i.e. at the core

Another possibility is that phase fluctuation away from the criticality may induce topological defects such as dislocations. Now, if one order nucleates as a topological defect in the other order, then it is likely to be created in the core of the ordered patch where the amplitude of the order parameter is strong and more or less uniform. Thus, the phase fluctuation governs the phase texture. We have shown schematically how such melting can occur in practice. Since we are sensitive to only the stripe order, both disordered and 'zigzag' phases appear invisible to the resonant X-ray scattering we employ (Fig. 4.22).

To carry out the investigation, we mounted the sample inside the scattering chamber with the c-axis perpendicular to the scattering plane. Then we tuned the polarization of the incoming beam to π polarization, and energy to the Fe-L edge at 707 eV. With the spin moments mostly pointing along the c-axis, a π polarized incoming beam is going to be highly sensitive to the spin component of the scattering factor. On the other hand, the σ polarized beam will be the least sensitive to the spin texture. We have illustrated the experimental setup in Fig. 4.23. Our $2th$ angle was 123° and $theta$ was 41° at such scattering angles the detector almost looks directly at the sample surface.

First, we show the temperature evolution of the integrated intensity of the antiferromagnetic peak. This data is continuously collected as we ramped the temperature while keeping the diffractometer aligned to the peak position of 41K. We integrated the intensity of the peak and showed it here as a function of the sample temperature. As expected, the antiferromagnetic order emerges at 47.5K and peaks around 41K where the 'zigzag' phase emerges. We see an immediate and sharp decrease in the 'stripe' phase intensity. We selected 5 temperatures including 41K to collect the speckle pattern as marked in the figure and elaborated below. Contrasting with the neutron experiment, we note that the intensity of the 'stripe' phase decreases significantly. This could be explained considering that the neutron scatters from the entire sample whereas we are scattering only from a region of $\sim$10μm (Fig. 4.24).

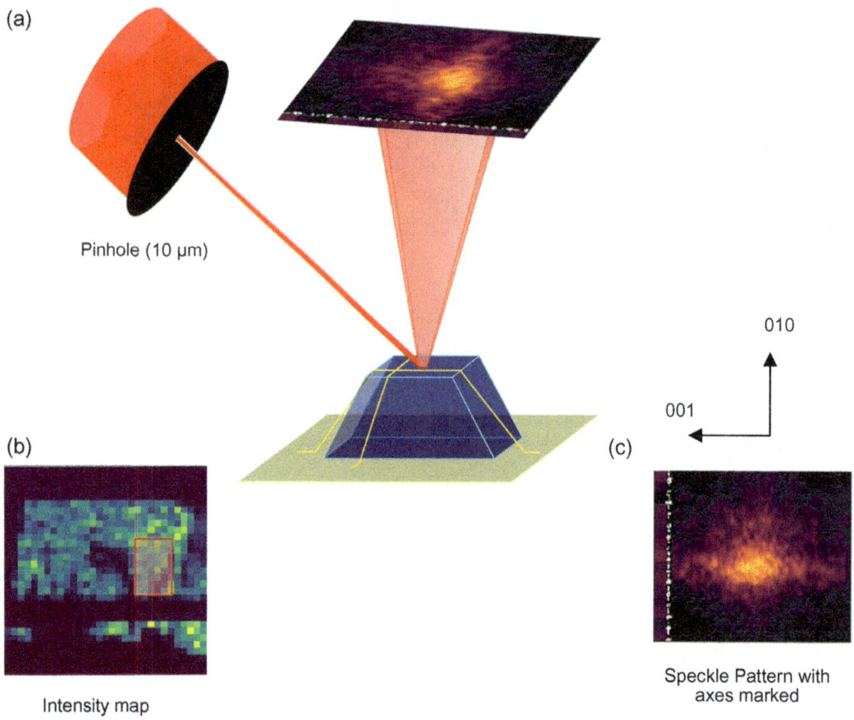

Fig. 4.23 Details of the experimental schematic a The schematic shows the pinhole selecting a small section of the beam incident on the sample. **b** Shows the integrated intensity of the peak on different part of the sample collected by rastering the beam. We investigated the red patch with coherent resonant soft X-ray scattering as we cycled the temperature. **c** The speckle pattern at 41K with the crystallographic axes marked

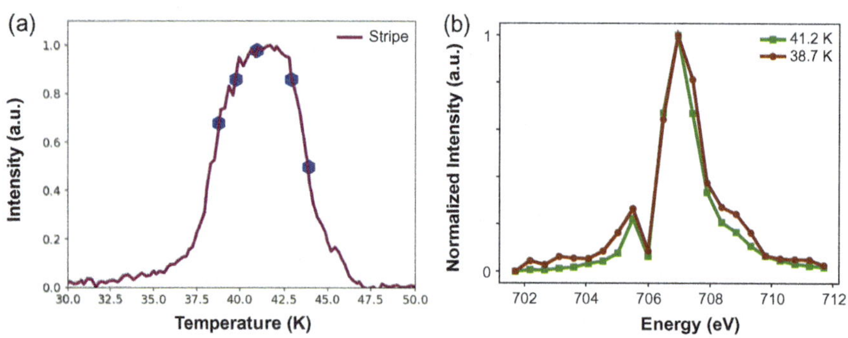

Fig. 4.24 Temperature and energy dependence: a Incoherent data on the 'stripe' peak showing temperature evolution of the intensity integrated over the peak, and **b** energy dependence from fixed-q energy scan on the peak position at two different temperatures, fully 'stripe' (41.2 K) and competing with 'zigzag' phase (38.7 K). The similarity indicates identical nearest neighbor correlation

From the energy response, we note that the peak position does not change with the 'stripe' phase melting under competition. We set our incoming beam energy to the peak of the energy response at 707 eV. We note a slight change in the energy response below 41 K, however, the overall peak shape remains similar highlighting an identical electronic correlation at the scale of nearest neighbor physics of the Fe atoms. The energy resolution of the incoming beam focusing optics is about 0.1 eV setting the longitudinal coherence length on the order of $1\,\mu m$ which is considerably higher than the penetration depth of the soft X-rays at the resonance condition ($\sim 80\,nm$). Thus, the entire scattering volume scatters coherently during the measurement we are about to discuss.

As seen in the figure above, the magnetic peak as observed in the fast CCD detector exhibits complex speckle patterns: a fingerprint of the spatial texture of the 'stripe' phase. At a given position, we captured the speckle pattern as we changed the temperature. However, this task was far from trivial and required us to use the gold wires as a homing device at each temperature, as discussed below.

4.4.2 Ensuring Positional Accuracy

As we change the temperature, the cryostat elongates. The motion is reflected as a change in the area illuminated under the beam. This potentially jeopardizes the experimental accuracy. Aligning ourselves at every temperature to the gold wires helps us re-home to the original position before changing the temperature. Two orthogonally oriented gold wires serve as two axes of a 2D coordinate system. Using the edge of the gold wires we aligned ourselves accurate up to the size of the beamspot of $\sim 10\,\mu m$ determined by the pinhole diameter of the same size.

We illustrate the experimental procedure of recalibration of the origin of the sample positioning coordinates under temperature change in the following figure. Each pixel in the intensity map is the integrated intensity on the 2D detector. It is easy to identify the gold wires as the linear dark regions. Right next to the dark horizontal line from the gold wire, we mark the region we explored to track the speckle pattern under temperature evolution with a translucent red patch (Fig. 4.25).

After setting up the origin, we moved to the center of the region by moving the sample motors. Then we waited long enough for the system (sample, sample motors, and the cryostat) to reach equilibrium, and did not move anything on the diffractometer after that. Then we moved only the pinhole using a nano-positioner to scan the beam on the sample. During the mapping of the sample using the nano-positioners of the pinhole we took $8\,\mu m$ steps. To ensure sub-beamspot size accuracy of positioning between two temperatures, we utilized features on the sample such as the one next to the region investigated. This feature had a single pixel giving off intensity enclosed by dark regions. Thus, it served as a nice homing beacon.

Finally, to ensure that we are measuring the same position at the same temperature, we also mapped the identical region using σ polarization. Unlike the π polarization, this is dominated by charge scattering. Thus the speckle pattern reflects structural

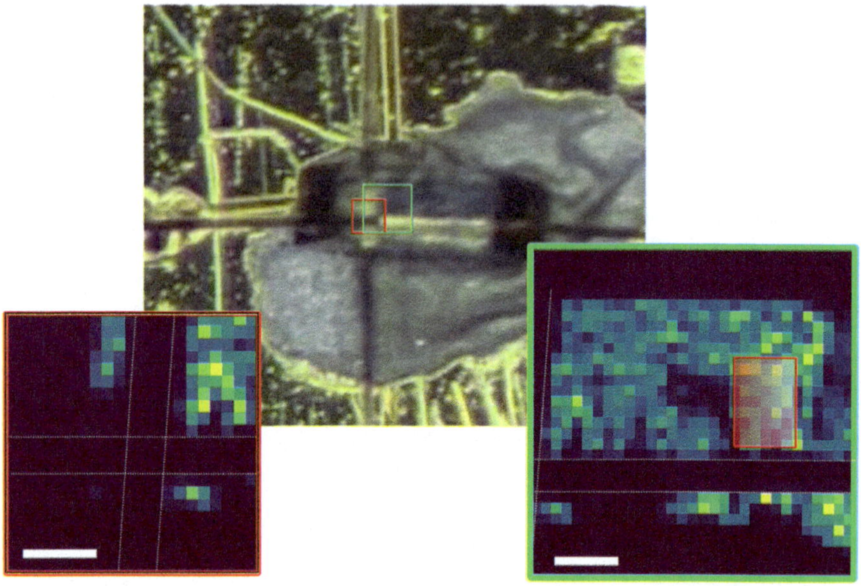

Fig. 4.25 Fiducial finding for spatial correspondence: Here we show how easily we could correlate the intensity maps of the antiferromagnetic peak with the shadow from the gold wires. The two instances of the nano-maps are shown. Each pixel is 10 μm and the scale bars are approximately 50 μm. Each maps are overlayed on an optical image for correspondence

features such as cracks or distortions on the surface and does not change with temperature. We used the speckle pattern generated from σ polarized incoming beam as the final check on our methodology to go to the original position. Figure 4.26 shows the speckle pattern from σ polarized incoming beam at different temperatures we explored. The 'streaks' coming off from the peak are fingerprints of the structural distortions present at that location. We consider this as a convincing proof of location match between different nano map measurements at different temperatures.

As an aside, we would like to draw the readers' attention to the fact that we could see a (0.5, 0, 0) peak with σ polarization, which is a forbidden Bragg peak. It is an allowed charge order peak for 'zigzag' phase. However, for under-intercalated samples that host the 'stripe' AFM phase, we observe a temperature-independent distortion peak at this location. Furthermore, this soft X-ray scattering beamline has a second harmonic contamination to the incoming energy. That may be responsible for a weak Bragg peak at this diffractometer position. However, as the 'zigzag' phase sets in the peak intensity is supposed to increase with lowering of temperature which we did not observe. This might be another piece of evidence that the charge-order doesn't form on Fe-atoms.

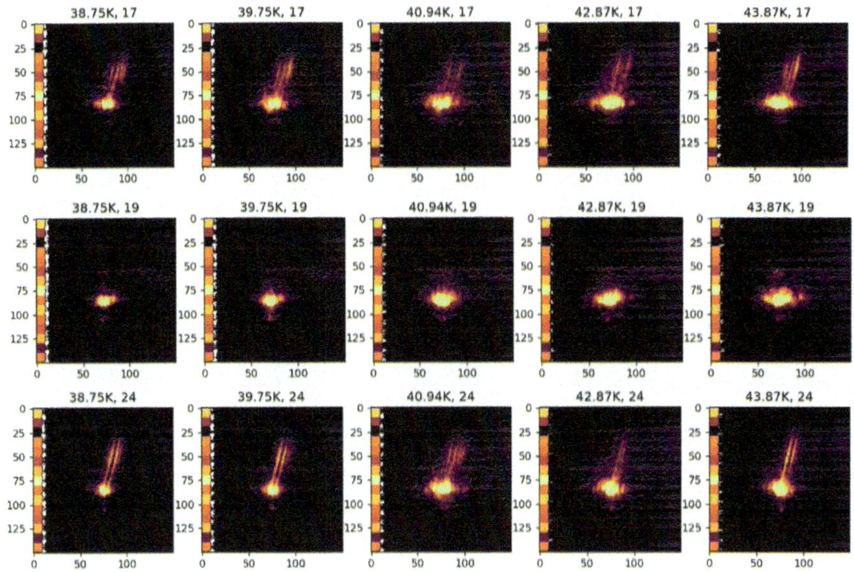

Fig. 4.26 Speckle pattern at horizontal polarization: Since the horizontal polarization is not sensitive to the spin moments the intensity pattern serves as a fingerprint of the position. Here we show horizontal intensity pattern for three such positions (indexed on top of each panel) showing the similarity of intensity pattern for every temperature visited on that spot (also mentioned on top of each panel). This intensity patterns acted as a sanity check on securing the positional accuracy

4.4.3 Analysis of Speckle Pattern

Now that we have established that we were certain of collecting the speckle pattern from the identical location to an accuracy of 8μm. We move on to discuss the data. In the Fig. 4.27 we show the π polarized speckle pattern of the 'stripe' phase texture. We have collected speckle patterns from a 11×11 grid. We ensured the data quality of each point and disregarded points that showed low intensity to discern the speckle pattern. We first note that the contrast is quite low indicating a highly incoherent background, either due to fast dynamics or a large fluorescence background. Our preliminary X-ray Photon Correlation Spectroscopy measurement indicated a slow fluctuation, convincing us of the latter as the reason.

We had more than 42 positions where the intensity was good enough to ensure high-quality speckle pattern analysis. As we discussed in the introductory chapters, for a complex speckle pattern with low contrast, we resort to correlation analysis for understanding of the speckle pattern. We show below the speckle pattern of a representative position and note the evolution of the speckle pattern at the temperatures 38.75 K, 39. 75K, 41 K, 42.87 K, and 43.87 K: two above and two below the peak saturation of the 'stripe' phase intensity.

To analyze the data, we follow the procedure laid out by, where we first evaluate the autocorrelation function of each pattern. Each auto-correlation pattern is a two-

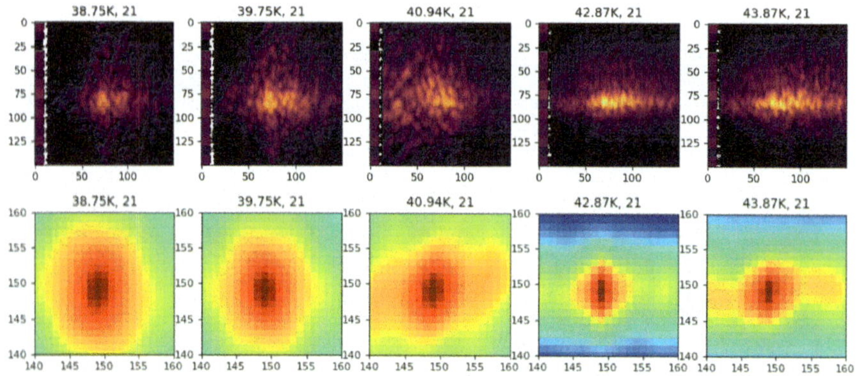

Fig. 4.27 Speckle pattern with vertical polarization and auto-correlation: (top row) shows that selecting vertical polarization for the incoming beam makes us sensitive to the antiferromagnetic order and hence the speckle pattern is a signature of the texture of the 'stripe' phase for every temperature sampled and mentioned on top of each panel. (bottom row) on the other hand measures the auto-correlation to extract the average speckle size which we use later to extract the incoherent background. We also mention the positional index (21) assigned following the above discussion

dimensional intensity distribution twice the size of the region of interest from the detector we selected. At the center of the intensity distribution lies a sharp peak on top of a gaussian background. This peak is the coherent contribution to the auto-correlation intensity and its size tells us the average speckle size. The extracted average speckle size in pixel units are 4×8, 4×8, 4×8, 2×4, and 3×4 in pixel units. This sharp peak sits on top of a broad Gaussian peak which comes from the incoherent part of the signal. We also note that the anisotropy of the correlation peak of the coherent portion is similar to the anisotropy of the individual speckle spots observed in the detector image.

Next, we utilize the average speckle size to evaluate the incoherent background from the speckle pattern. We subtracted the incoherent background from the raw data to get the coherent scattering data. We used a Gaussian filter of the same size as the average speckle size to wash out the intensity modulation. This gives us an incoherent background as shown in the Fig. 4.28.

However, incoherent data holds valuable information, such as the correlation length of the order. We note that the vertical direction on the detector is mostly parallel to the (010) direction, whereas the horizontal direction is parallel to the (001) direction of the crystal reciprocal lattice. We took cuts of the intensity along the horizontal and vertical directions and subsequently fitted them with a Gaussian signal. We observe that along the vertical direction, we have a single Gaussian intensity distribution. However, along the horizontal direction, we have two intensity maxima, which we fit with a single Gaussian too. Our fit matches with the average of both intensities to extract the average correlation length along the direction. With their reciprocal lattice marked, we conclude that the out-of-plane correlation length increases slightly as the stripe phase melts. But, the correlation length along the 010 direction decreases

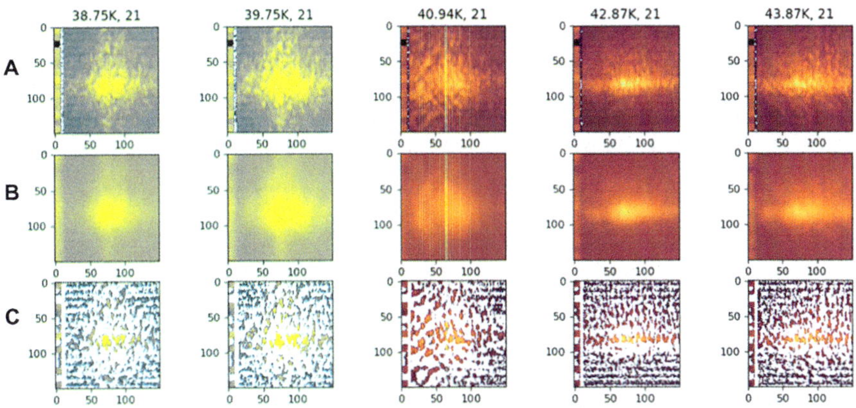

Fig. 4.28 Incoherent background and coherent signal extraction: From row A we evaluate the auto-correlation function to extract the average speckle size which we use as the Gaussian filter size to find the incoherent background shown in row B. Finally the row C is extracted by subtracting each column of B with respective A panel. This row then shows entirely coherent data

approximately by a factor of 2 right at the boundary, where the 'zigzag' phase sets in. We show in the inset the directions on the detector. The figure on the left is used to show that the vertical direction of the detector is mostly parallel to the 010 direction, whereas the horizontal direction which is parallel to the 001 direction cannot be seen in the figure. The decrease in correlation length by a factor of 2 saturates as we further lower the temperature. However, we notice that the overall peak intensity decreases significantly. We interpret this result as disorderliness in correlation central to the amplitude-correlated domain structures. This is our first hint at solving the problem of phase competition (Fig. 4.29).

Once we have filtered out the incoherent signal we employ cross-correlation between the speckle patterns of two different temperatures. The cross-correlation between the speckle pattern taken at T_i and T_j is called as C_{ij}, and defines as follows:

$$C_{ij} = \sum_{m=1}^{N} \sum_{n=1}^{N} A(m, n) \times B(m - i, n - j) \tag{4.6}$$

Here A and B are $N \times N$ matrices of the intensity distribution of the speckle pattern for T_i and T_j, respectively. Physically, this is equivalent to shifting one speckle pattern on the other by (i, j) and multiplying the overlapping pixels together. In this way, we extract the intensity-intensity correlation between two temperatures, resulting in a four-point correlation function. The cross-correlation matrix is a symmetric matrix with diagonal elements as the auto-correlation of each image (Fig. 4.30).

Interestingly, the cross-correlation also shows complex patterns. However, we will focus on the central intensity peak which captures the similarity between the two images. We are following here the methodology for two-time correlation evaluation

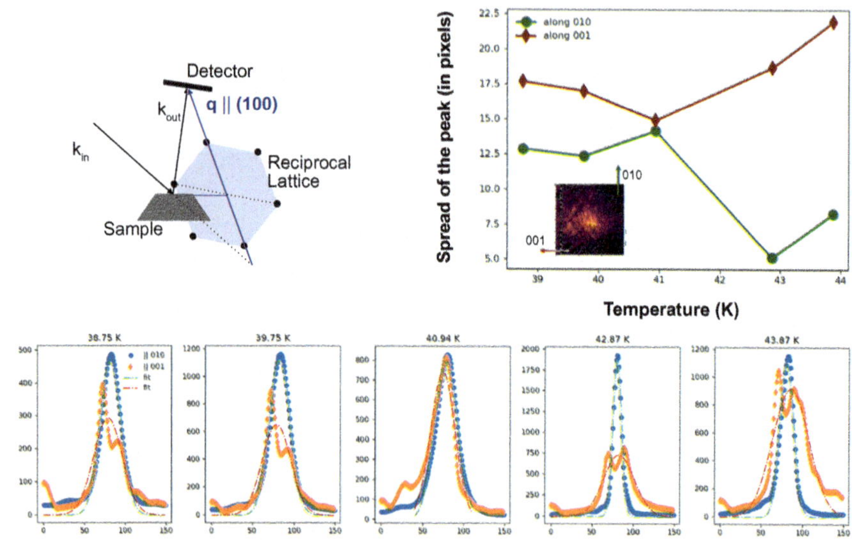

Fig. 4.29 Evolution of FWHM in the incoherent envelope on the speckle patterned 'stripe' peak. The crystallographic direction on the detector is extracted from the figure on the top-left. The FWHM of the peak along these high symmetry direction is extracted and fitted with Gaussian to find the evolution of the FWHM. Notably, we see that along (010) the FWHM doubles at the 'stripe' to 'zigzag' cross-over

for X-ray photon correlation spectroscopy. Instead of two-'time' we are finding the four-point correlation between two temperatures.

Finally, we evaluated normalized cross-correlation between two images which serves as a similarity index of the speckle pattern. This is defined as—

$$\rho_{ij} = \frac{\sum_{x,y} C_{ij}}{\sqrt{\sum_{x,y} C_{ii} \sum_{x,y} C_{jj}}} \tag{4.7}$$

This is a number between 0 to 1. If the two images are identical ,the similarity index will be 1, whereas a completely anti-correlated image will yield 0. We carried out the sum in an 8×8 pixelated region centrally. In Fig. 4.31 we show the similarity index matrix or two-temperature correlation for the 5 temperature points measured.

We observe a block diagonal form where the two low-temperature speckle patterns match highly with each other and the two high-temperature speckle patterns match. However, neither of these sets match the 41K. Since the low-temperature speckle patterns do not match with the high-temperature speckle pattern, we have a clean first evidence that the melting of the antiferromagnetic 'stripe' phase due to thermal fluctuation is different from the melting due to phase competition with the 'zigzag' phase.

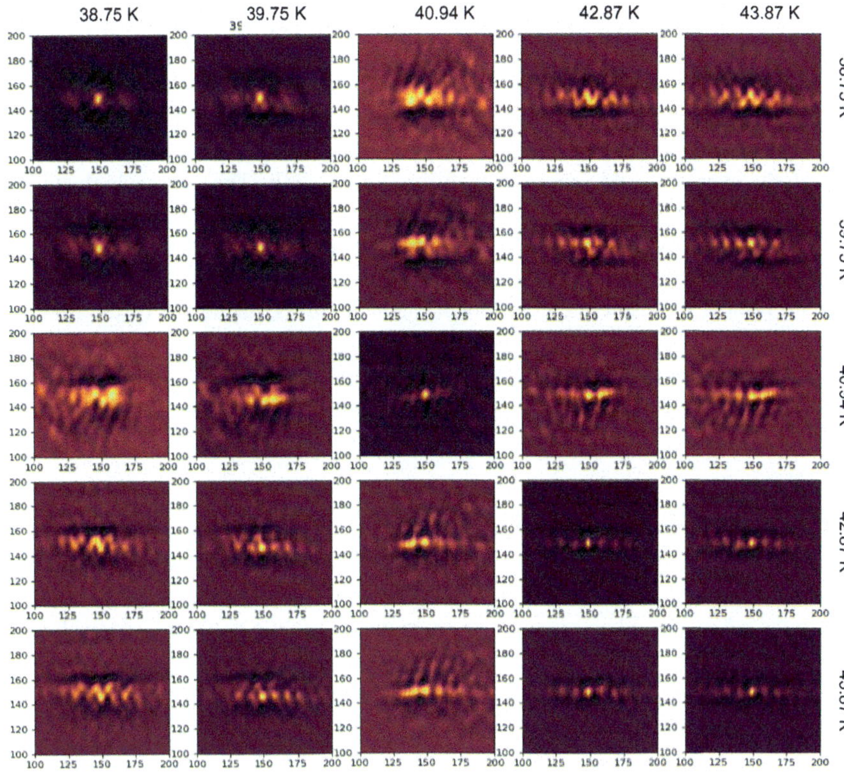

Fig. 4.30 Cross correlation evaluation between two temperature speckle patterns: we evaluated the two temperature cross correlation by convolution one speckle pattern with another. The diagonal entries are auto-correlation matrices. This is used to extract the similarity index

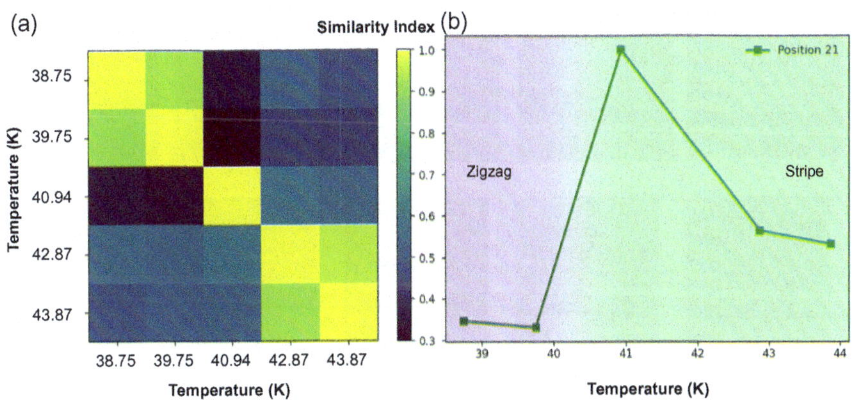

Fig. 4.31 Similarity index evaluation: a Similarity index between two temperature speckle pattern **b** similarity of different temperature speckle pattern with 41.2 K speckle pattern. At the 'stripe' to 'zigzag' crossover there is a sharp fall of correlation

The mismatch with the 41 K speckle pattern is elaborately shown in the right panel. This plot can be thought of as a line cut through the central row or central column of the two-dimensional matrix shown on the left. Contrary to the high-temperature side, where the similarity drops gradually, it drops rapidly on the low-temperature side. The change of intensity is relatively similar for these two temperatures, even though the overall speckle correlation differs significantly. The relative high similarity between two temperatures can happen if the ordered regions grow or melt from the boundary. Since the speckle pattern observed after scattering is the power spectrum of the Fourier transform, a boundary expansion or shrink operation on each domain retains the principle component of the spatial distribution of the domains and thus the principle component of the speckle pattern. This is observed for the two end-points on the temperature axis in each side. To illustrate how two speckle patterns may match while the overall intensity decreases, we resort to modeling.

4.4.4 Modeling the Melting

Here we show the cross-correlation among the speckle pattern generated from two situations we discussed an order phase patch would melt: **a** melting from the boundary, and **b** melting from the core.

On the top row (A1–A5) of Fig. 4.32, we have a circular beam spot, illuminating a region on the sample with an arbitrary representative heterogeneous distribution of

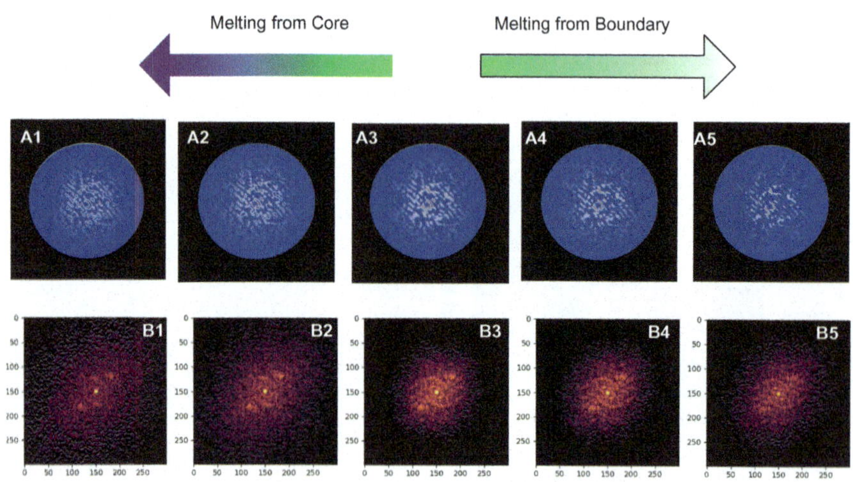

Fig. 4.32 Modeling of the observed speckle pattern: (A1–A5) Shows real space texture evolution under the circular gaussian coherent beam-spot. The texture of A3 is modelled to melt from the outer boundary in A4 and A5, and from the core in A2 and A1. (B1–B5) shows corresponding reciprocal space speckle pattern evaluated by taking Fourier transform of the corresponding real space texture

antiferromagnetic phase. The bottom row (B1–B5) is the amplitude Fourier transform of this illuminated domain distribution, which looks like a complex speckle pattern. We considered a Gaussian beam illumination and thus multiplied the domain distribution with a Gaussian envelope to generate the real space scenario similar to what we had during the experiment. For the reciprocal space data, we just took the power spectrum of the Fourier transform of the domain distribution. Let us navigate this figure from the middle column (A3 and B3) which represents a situation in the phase transition where the stripe phase is at its maximum strength.

From this condition, we consider the two pathways of melting. The melting from the core begins from the second column from the left and progresses towards the left (A2 to A1). Whereas, the melting from the boundary scenario is illustrated on the right two panels (A4 to A5). At the first glance, we note that the Gaussian envelope of the intensity distribution in reciprocal space for the first two panels (B1 and B2) shows an increase in FWHM, denoting a sudden decrease in the correlation length. This sudden decrease in correlation can also be seen in our experimental data between 42.87 K and 40.94 K suggesting a similar stripe phase mesoscopic texture evolution between those two temperatures. On the other hand, melting from the boundary gradually decreases the amplitude correlation length. We first focus on the melting evolution from A3 to A5. The overall intensity decreases as the scattering volume decreases (Fig. 4.33).

Next, we focus on the fine details. The speckle patterns on the Gaussian envelope were extracted similarly and then evaluated with Eq. 5.3. In Fig. 4.34 we show the speckle pattern after the incoherent background subtraction and zoomed into a region of interest on the peak. The incoherent background comes from the average correlation length discussed above. The speckle pattern, on the other hand, tells us about the domain organization. Before quantifying the similarity index of the speckle pattern, we qualitatively discuss this point.

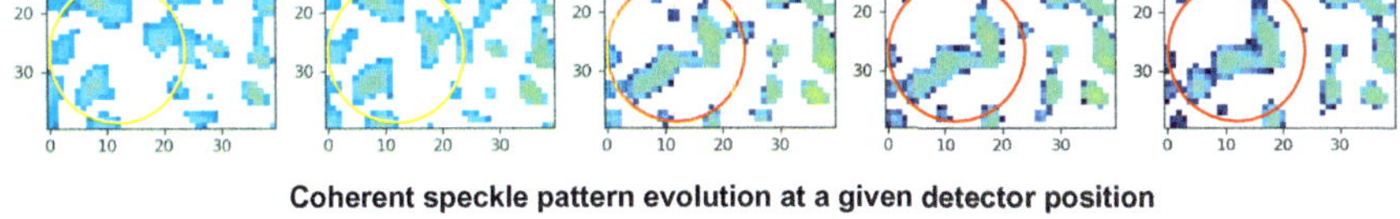

Fig. 4.33 Zoomed-in view to observe changes in speckle pattern in modeling: (C1–C5) shows a zoomed in section from (B1–B5). The similarity of speckle pattern for example can be understood looking inside the red-circle marked area. From C3 to C2 the speckle pattern is extremely dissimilar as melting from core happens

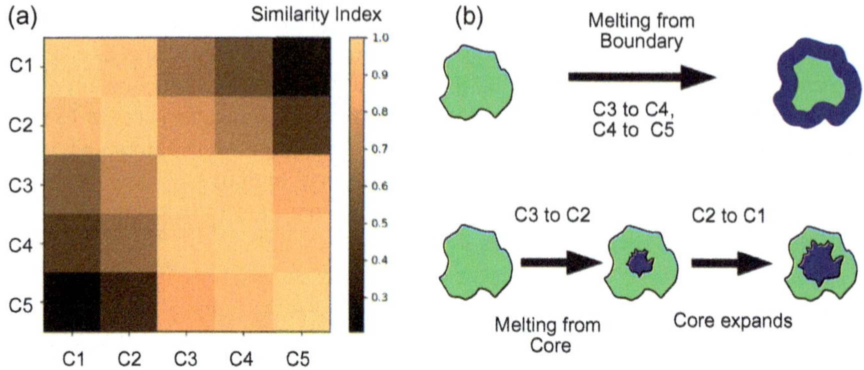

Fig. 4.34 Evaluation of similarity index for the modeling simulation: a Similarity index for the speckle patterns evaluated. They behave similar to experimental data. **b** Shows schematic of melting of arbitrary ‘stripe‘ phase domains for each step of modeling

If we follow the speckle pattern from C3 to C5 we identify large structures that retain position, intensity, and shape similarity among the panels. Low-intensity small structures are the first ones to become dissimilar. The similarity index between two such speckle patterns will be found to be high. A similar evolution of the speckle pattern is observed between C2 to C1. Thus we argue that in later stages of melting from the core, the evolution of the speckle pattern indicates a melting from the boundary for the annular domain areas, which hosts a phase decorrelated core in the middle. However, the speckle pattern completely reorganizes between C2 and C3, which indicates that the nucleation phase of phase decorrelated core will completely reorganize the speckle pattern. The similarity index between two such speckle pattern will be low. Once the cores are established and their further evolution involves growing in volume, the similarity index between two such speckle patterns will be high.

4.5 Discussion

In light of the previous argument, we revisit the similarity index matrix from our data. The block diagonal structure now can be understood as temperature regions where melting from the boundary is the dominant mode of melting. So, from 39.75 K to 38.75 K, and from 42.87 K to 43.87 K the stripe phase domains melt from the core boundary and domain boundary, respectively. However, the similarity index among the speckle patterns of 42.87 K, 40.94 K, and 39.75 K are quite low. We conclude that between 42.87 K to 40.94 K the core regions of the stripe domains start to disappear. We consider that the cores are replaced by the zigzag phase, since the low-temperature melting of the ‘stripe’ phase is associated with the phase competition with the ‘zigzag’

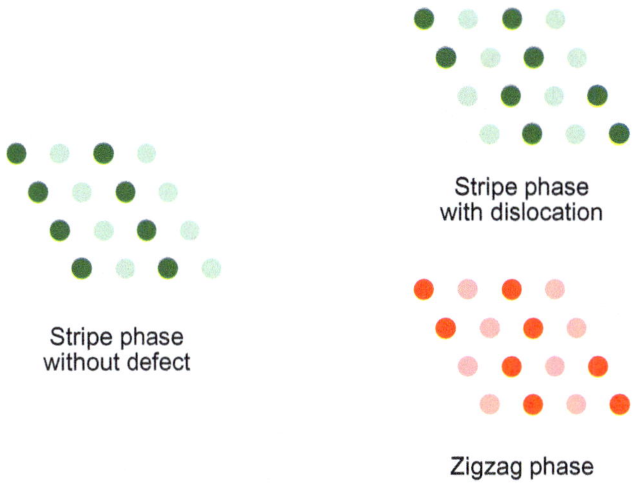

Fig. 4.35 Correlation between stripe phase and zigzag phase: Here we show that for a given domain orientation a stripe phase with dislocation looks identical to a zigzag phase spin arrangement

phase. Between 40.94 K and 39.75 K, the 'zigzag' phase continues to populate the real space by further nucleation and the growth of existing ones.

Following the methodology discussed, we analyzed all the positions in question to gain statistical significance. This leads us to a conviction that the observed behavior of melting is not special for the representative case. It happens all over the region we investigated.

Thus, we conclude that the melting due to the 'zigzag' phase below 41K happens from the core of the stripe patch, as if the 'zigzag' phase is nucleated as a topological defect of the 'stripe' phase. With the help of coherent resonant soft X-ray scattering, we have found the key relationship between the two phases. Following our understanding, first, we show below in Fig. 4.35 schematically how the 'zigzag' phase can be viewed as a topological defect for the 'stripe' phase. Then, we discuss a phenomenological theory that encompasses the essential physics of the system.

4.5.1 *Looking Through McMillan-Ginzburg-Landau Theory*

Figure 4.35 captures the relationship between the order parameters schematically. There are three equivalent wave vectors for the stripe phase related by a 120° where we have shown the relationship for only one of them. Let us consider the magnetic moment at a site $\mathbf{r_i}$ is

$$\mathbf{M}(\mathbf{r_i}) = e^{\iota \mathbf{q}_{\text{stripe}} \cdot \mathbf{r_i}} M,$$

where $\mathbf{q_{stripe}}$ is the reciprocal lattice vector of the magnetic order and M is the order parameter. Now, the topological defect in this field can be defined similarly to Burger's vector as

$$\oint_C \nabla \mathbf{M} \cdot d\mathbf{r}$$

taken around a loop C. If C does not enclose a dislocation defect in the stripe phase, the path runs over lattice spacing to yield 0. If, however, it encircles a defect, the integration yields $\iota 2\pi M$, which is equivalent to the winding number. On the other hand, the same path of integration is followed in the zigzag phase where

$$\mathbf{M}(\mathbf{r_i}) = e^{\iota \mathbf{q_{zigzag}} \cdot \mathbf{r_i}} M,$$

the dot product for the entire loop cancels out and we find the order is defect-free. Since the two antiferromagnetic phases are very close in energy space and are related by a dislocation defect, a phase fluctuation may result in nucleating the zigzag phase in the stripe phase as a topological defect of the latter.

To capture this into a phenomenological theory, we borrow inspirations from the McMillan-Ginzburg-Landau model of commensurate to incommensurate charge density wave (CDW) transition through discommensuration [17, 18] . The second-order transition from commensurate to incommensurate is seen on transition metal dichalcogenide 2H-$TaSe_2$. The discommensuration is defined as the phase of the complex order parameter of the CDW commensurate phase which differs from $2\pi/3$ effectively unlocking the CDW from the crystal lattice. The discommensuration thus is a phase defect in the commensurate CDW field. Following the solution of the free energy minimization for a constant amplitude but varying phase of the complex order parameter, the authors show that discommensuration nucleates well separated from each other and eventually span the space with decreasing temperature.

Let us redefine the magnetic moment at a site as

$$\mathbf{M}(\mathbf{r_i}) = e^{\iota \mathbf{q_{stripe}} \cdot \mathbf{r_i}} \psi,$$

where ψ is a complex order parameter defined in real space as $\psi(\mathbf{r}) = \phi(\mathbf{r})e^{\iota\theta(\mathbf{r})}$. The phase quantifies the deviation from an order parameter of a certain wave vector. Then the McMillan free energy functional can be written as

$$f = \tau|\psi^2| + \xi^2|(\iota\nabla + \mathbf{q}_\delta)\psi|^2 - \frac{E}{2}(\psi^2 + \psi^{*2}) + G|\psi^4| \tag{4.8}$$

where τ is the reduced temperature and G,E> 0. ξ is the characteristics length scale of correlation and $\mathbf{q}_\delta$ is a reciprocal space vector such that

$$\mathbf{q_{stripe}} + \mathbf{q}_\delta = \mathbf{q_{zigzag}}.$$

We did not consider any coupling to external fields or the charge order term to keep the equation to the bare minimum terms necessary to explain the phenomena. Including them does not change the fundamental structure of the argument presented here. Furthermore, this free energy density is only for a single domain where one of the order parameters manifests out of three possibilities, ψ_1, ψ_2, and ψ_3 rotated 120° to each other. A full free energy density will include such terms for every ψ_i and the symmetry-allowed coupling terms. Now, replacing ψ with $\phi e^{\iota\theta}$ in the Eq. 4.8 we get:

$$f = \tau\phi^2 + \xi^2[(\nabla\phi)^2 + \phi^2(\mathbf{q}_\delta - \nabla\theta)^2] - \frac{E}{2}\phi^2 cos(2\theta) + G\phi^4 \tag{4.9}$$

The reduced temperature term decides if the ground state energies will have an order with non-zero ϕ, however, the wave vector for the order parameter is decided by the subsequent terms. The second term minimizes the free energy when $\nabla\theta = \mathbf{q}_\delta$ or, $e^{\iota\theta} = e^{\iota\mathbf{q}_\delta\cdot\mathbf{r}}$. On the other hand, the third term minimizes the free energy the most when $\theta = n\pi$ with n as an integer. Thus, the term does not alter the ordering vector of the moments but changes the overall phase by ± 1. Through the free energy density description we have incorporated the emergence of the 'zigzag' phase and 'stripe' phase as per the phase(θ) term of the complex order parameter field. In the limit when $T << T_C$ and the order parameter is established to a constant, the free energy density is dictated by the phase term as we considered previously.

The Euler-Lagrange equation for the phase term can be found from the minimization of the functional -

$$\frac{\delta f}{\delta\theta} = 0 = \nabla^2\theta - \frac{2E}{\xi^2} sin(2\theta) \tag{4.10}$$

This is in the form of the Sine-Gordon equation, which allows soliton-like solutions for the phase. From a complex order parameter field following the McMillan-Ginzburg-Landau energy functional, we have found that a significant value of E favors the formation of the 'stripe' phase along with soliton-like solution for the phase term(θ) of the complex order parameter field. We have also identified such a soliton-like solution for the 'stripe' phase as the 'zigzag' phase. This explains the nucleation of the 'zigzag' phase in the 'stripe' phase as soon as the amplitude fluctuations(ϕ) are diminished and phase (θ) fluctuation becomes the main driver of the physics.

Solving the Sine-Gordon equation, we come to the solution for the phase(θ) term as it evolves over space.

$$\theta(r) = C * tan^{-1}(e^{\pm\frac{2\sqrt{E}}{\xi}r}), \tag{4.11}$$

where r is the distance along a direction in the lattice plane perpendicular to the c-direction and C is some constant. We show in Fig. 4.36 how the solution varies for $\xi = 2$ and $E = 0.1, 0.7, 4,$ and 25. We see that a high E value stabilizes θ to integer multiples of π, which allows the 'stripe' phase. As a solution of the Sine-

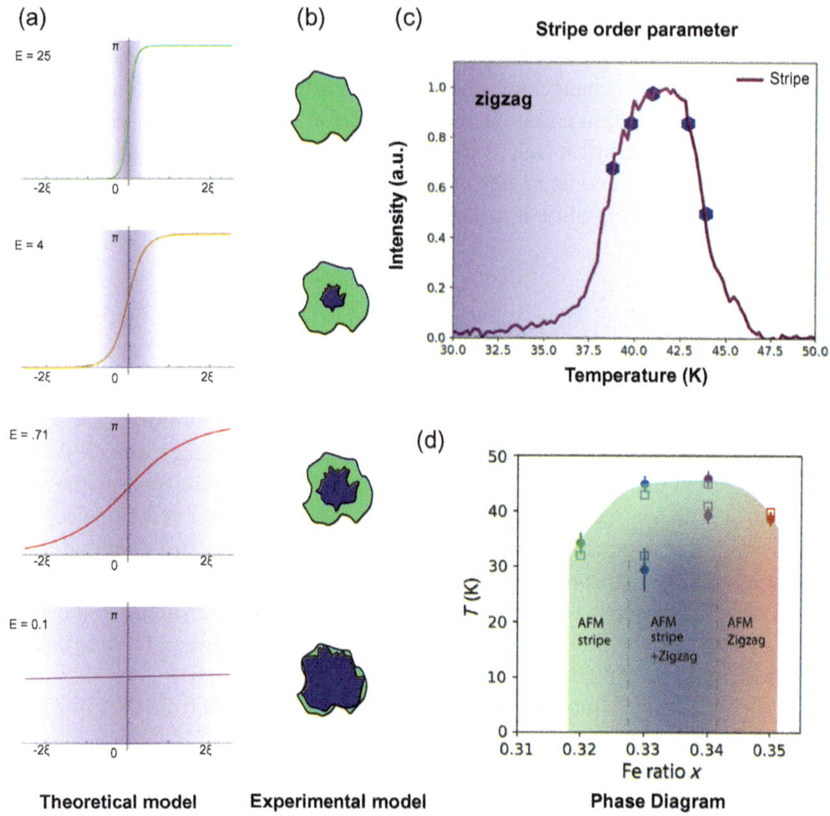

Fig. 4.36 Conclusion of melting of the stripe phase and comparison of theoretical and experimental models: a purple shaded region is the competing 'zigzag' phase which nucleates and grows only varying the coupling parameter E, representing magneto-elastic lock-in energy. **b** Schematically represented evolution of a stripe domain (in green) as 'zigzag' phase competes. **c** Temperature evolution of the 'stripe' phase observed with resonant soft X-ray scattering **d** phase diagram of the material repeated for context

Gordon equation, we observe the 'zigzag' phase emerge as a defect. As the value of E decreases, this phase increases in expanse and wipes the 'stripe' phase out.

We conclude that we have found a qualitative similarity between the phenomenological model and the experimental model based on the observation. McMillan-Ginzburg-Landau functional captures the essential physics of the system that we have realized from the observation of the coherent resonant scattering outcome. Tracing the evolution of the speckle pattern during the melting of the stripe phase during competition and thermal melting showed the intricate interrelation between the two phases. The E term in the functional is dimensionless and is explained by McMillan as the phase coupling to the structure in this model, termed as lock-in energy. This is possibly the manifestation of the magneto-elastic coupling discussed in relation to the concomitant charge order in this material. We know from the phase diagram that the term can be modulated by intercalation ratio. For a lower intercalation ratio than the stoichiometric one, it results in a higher value of E realizing the 'stripe' phase as the low-temperature ground state. With increasing intercalation, the value switches to a lower value and we observe the 'zigzag' phase. Around the stoichiometric ratio, we observe a temperature-dependent evolution that nucleates the 'zigzag' phase inside the 'stripe' phase and allows it to expand with decreasing temperature. The temperature dependence of the E term may result from the small thermal expansion in this strongly correlated system.

4.5.2 Outlook

Physics of strongly correlated magnetic systems is fascinating. In this chapter, we have explored one such system. We studied the evolution of the antiferromagnetic texture to identify the physics using coherent resonant soft X-ray scattering. In the outlook section, we discuss a few loose ends and how we are approaching to resolve them.

Charge order in under-intercalated sample Even though the Landau theory term indicated the existence of a charge-ordered phase concomitant with the antiferromagnetic order in the under-intercalated sample, we have not directly observed it. An indirect way to observe could come from exploring coherent hard X-ray scattering. Critical slowing down during phase transition is a universal phenomenon in continuous phase transition. We can study the fluctuation dynamics of the structural domains and domain walls as the sample undergoes a phase transition using X-ray Photon Correlation Spectroscopy (XPCS). The time scale of the dynamics tells us the energy separation between two fluctuating states of domain configuration. Thus, by tuning to a diffuse scattering in the tail of a Bragg peak we can monitor the time dynamics at different temperatures. A critical slowing down will indicate the existence of an emergent charge order. We can simultaneously study the inelastic X-ray scattering which will show the softening of the phonon mode associated with the charge order formation.

Short range interaction: Coherent scattering explored the long wavelength physics of the material. To explore the nearest neighbor coupling term, we can perform a resonant inelastic soft X-ray scattering experiment. Furthermore, we can study the absorption spectroscopy of the sample to understand the unoccupied electron levels.

In-operando measurements: Now that we have an understanding of the phase evolution in the sample in thermal equilibrium we can extend it to studying the non-equilibrium physics during resistive switching phenomena which originally got us interested. In such a measurement, we can use Dark Field X-ray Microscopy (DFXM) measurement to image the melting of the associated charge order phase and re-emergence through the switching cycle. We can also study the domain fluctuation dynamics on the charge order through XPCS as we vary the switching frequency.

4.6 Conclusion

We observed a remarkable antiferromagnetic phase in intercalated NbS_2 samples. The antiferromagnetism couples to a charge-ordering phenomenon. The origin of the charge order remains unknown. However, we have resolved the relation between two magnetic phases that show up in this material with coherent resonant soft X-ray scattering. We can clearly distinguish between the melting of the stripe phase under thermal fluctuation and under phase competition. Furthermore, the phase competition indicates an interesting relationship between the two magnetic phases where one phase is a topological defect of the other phase. From this relation, we developed a phenomenological model based on McMillan-Ginzburg-Landau model which qualitatively agrees with the experimental observation of the 'stripe' phase melting under phase competition. The methodology employed using resonant coherent soft X-ray scattering can be employed elsewhere similarly, where an order parameter of a phase melts in competition with another phase.

References

1. Nair NL, Maniv E, John C, Doyle S, Orenstein J, Analytis JG (2020) Electrical switching in a magnetically intercalated transition metal dichalcogenide. Nat. Mater. **19**(2), 153–157
2. Wu S, Xu Z, Haley SC, Weber SF, Acharya A, Maniv E, Qiu Y, Aczel AA, Settineri NS, Neaton JB et al (2022) Highly tunable magnetic phases in transition-metal dichalcogenide fe 1/3+ δ nbs 2. Phys Rev X 12(2):021003
3. Callen ER, Callen HB (1963) Static magnetoelastic coupling in cubic crystals. Phys Rev 129:578–593
4. Harris QJ, Feng Q, Lee YS, Kim Y-J, Birgeneau RJ, Ito A (1997) A synchrotron x-ray study of the phases and phase transitions in the mixed ising-xy magnet $Fe_xCo_{1-x}TiO_3$. Z Phys B: Condens Matter 102(2):163–182
5. Kiss T, Yokoya T, Chainani A, Shin S, Hanaguri T, Nohara M, Takagi H (2007) Charge-order-maximized momentum-dependent superconductivity. Nat Phys 3(10):720–725
6. Borisenko SV, Kordyuk AA, Zabolotnyy VB, Inosov DS, Evtushinsky D, Büchner B, Yaresko AN, Varykhalov A, Follath R, Eberhardt W, Patthey L, Berger H (2009) Two energy gaps and fermi-surface "arcs" in $nbse_2$. Phys Rev Lett 102:166402
7. Rahn DJ, Hellmann S, Kalläne M, Sohrt C, Kim TK, Kipp L, Rossnagel K (2012) Gaps and kinks in the electronic structure of the superconductor $2h$-$nbse_2$ from angle-resolved photoemission at 1 k. Phys Rev B 85:224532
8. Straub T, Finteis T, Claessen R, Steiner P, Hüfner S, Blaha P, Oglesby CS, Bucher E (1999) Charge-density-wave mechanism in $2H - nbse_2$: photoemission results. Phys Rev Lett 82:4504–4507
9. Rossnagel K, Seifarth O, Kipp L, Skibowski M, Voß D, Krüger P, Mazur A, Pollmann J (2001) Fermi surface of $2h - nbse_2$ and its implications on the charge-density-wave mechanism. Phys Rev B 64:235119
10. Valla T, Fedorov AV, Johnson PD, Glans P-A, McGuinness C, Smith KE, Andrei EY, Berger H (2004) Quasiparticle spectra, charge-density waves, superconductivity, and electron-phonon coupling in 2H-$nbse_2$. Phys Rev Lett 92:086401
11. Shen DW, Zhang Y, Yang LX, Wei J, Ou HW, Dong JK, Xie BP, He C, Zhao JF, Zhou B, Arita M, Shimada K, Namatame H, Taniguchi M, Shi J, Feng DL (2008) Primary role of the barely occupied states in the charge density wave formation of $nbse_2$. Phys Rev Lett 101:226406

12. Weber F, Rosenkranz S, Castellan J-P, Osborn R, Hott R, Heid R, Bohnen K-P, Egami T, Said AH, Reznik D (2011) Extended phonon collapse and the origin of the charge-density wave in $2h$-nbse$_2$. Phys Rev Lett 107:107403
13. Weber F, Hott R, Heid R, Lev LL, Caputo M, Schmitt T, Strocov VN (2018) Three-dimensional fermi surface of $2h - NbSe_2$: implications for the mechanism of charge density waves. Phys Rev B 97:235122
14. Hashimoto M, He R-H, Testaud JP, Meevasana W, Moore RG, Lu DH, Yoshida Y, Eisaki H, Devereaux TP, Hussain Z, Shen Z-X (2011) Reaffirming the $d_{x^2-y^2}$ superconducting gap using the autocorrelation angle-resolved photoemission spectroscopy of $bi_{1.5}pb_{0.55}sr_{1.6}la_{0.4}cuo_{6+\delta}$. Phys Rev Lett 106:167003 (2011)
15. Youbi ZE, Jung SW, Richter C, Hricovini K, Cacho C, Watson MD (2021) Fermiology and electron-phonon coupling in the $2h$ and $3r$ polytypes of Nbs$_2$. Phys Rev B,103:155105
16. Lee S, Collini J, Sun S X-L, Mitrano M, Guo X, Eckberg C, Paglione J, Fradkin E, Abbamonte P (2021) Multiple charge density waves and superconductivity nucleation at antiphase domain walls in the nematic pnictide ba 1- x sr x ni 2 as 2. Phys Rev Lett 127(2):027602
17. McMillan WL Theory of discommensurations and the commensurate-incommensurate charge-density-wave phase transition
18. Moura VN (2024) McMillan-Ginzburg-Landau theory of singularities and discommensurations in charge density wave states of transition metal dichalcogenides. Phys Rev B

Chapter 5
Beyond: Multiple Beam-Spot Coherent Scattering

Abstract This chapter serves as an outlook for the thesis, highlighting possibilities beyond current capabilities in coherent resonant X-ray scattering. We introduce a simple approach to generating multiple beamspots by modifying Fresnel zone plates, a standard optic for focusing coherent beams at nano-diffraction facilities. Several designs are presented along with detailed guidance. This framework illustrates how tailoring the incident beam can simplify experimental observations while also enhancing existing modalities. More importantly, it opens new opportunities for probing systems that exhibit flow—whether of mass, charge, spin, energy, or entropy—where the gradient between distinct beamspots becomes the key observable in the coherent scattering signal.

5.1 Introduction

With the ability of coherent resonant X-ray scattering laid out, we observe that it can be used to

- image first formed domains of antiferromagnetic order during nucleation of a first order phase transition
- image artificially fabricated lateral hetero-structures of mesoscale antiferromagnetic patterns in a paramagnetic background
- distinguish between the melting of an antiferromagnetic order to phase competition and thermal fluctuation and identify the competing phase as the topological defect of the antiferromagnetic order.

We want to highlight that, even though we showed the results for antiferromagnetic order, the method is not specific to antiferromagnetic order. Any electronic order such as charge order, density wave order, or orbital order is observable by the coherent soft-Xray resonant scattering. As long as the peak is broad in reciprocal space and we have enough contrast so that the Bragg peak hosts a speckle pattern, we can unearth information about the mesoscale structure and dynamics.

R. Basak, *Nucleation, Fabrication, and Competition in Antiferromagnetic Textures Studied with Coherent Resonant Soft X-ray Scattering*, Springer Theses,
https://doi.org/10.1007/978-3-032-19199-1_5

However, at the outset, the method appears to be complex. This is because of the complex speckle pattern associated which does not provide us with a direct real-space picture, but hides the information in the correlations as we saw in the previous chapter. At its core, the complexity rises from the multiplicity of the domains under the beam spot as shown in Chap. 3 [1]. Only in the dilute limit, does the speckle pattern become easy enough to 'guess' the spatial distribution of the domains correctly. Here we want to leverage this 'diluteness' aspect and show its potential to study the physics of systems under dynamic equilibrium.

Even if the domain distribution is not dilute, diluting the beam footprint on a random domain distribution will have the same simplifying effect on the speckle pattern. This is our key argument and is shown here schematically in Fig. 5.1 for a random domain distribution. As seen previously, the simplification offers the ability to extract the spatial distribution of domains from a single shot of the speckle pattern. Tracking the time evolution of the speckle pattern under experimental conditions will reveal the spatio-temporal structure of domains if and when it changes.

To help dilute the beam footprint we will use the modifications to the already employed X-ray optics in the scattering chambers of soft X-ray resonant coherent beamlines, specifically the Fresnel zone plate. All of the figures shown here are simulations to illustrate the point in question. Also, through the process, we will show the designs that have been developed to create multiple coherent beam spots of illumination. In the following section, we will discuss this in detail. Once we have established the design schema for the modification we will discuss science questions that can be addressed with it.

5.2 Results: Beamspot Engineering

Creating a coherent beam spot from an extended source requires focusing. There are multiple ways to focus X-rays such as Kirkpatrick-Baez mirrors, kinoform lens, compound refractive lens, and Fresnel zone plates. We will briefly mention the others before we focus on the main element of discussion: Fresnel zone plate. Kirkpatrick-Baez mirrors are probably the most widely used method to image the source at a desired focal spot. This is an assembly of two orthogonal cylindrical mirrors set in grazing reflection conditions. Apart from reflective optics, refractive optics are also used to focus X-rays, known as compound refractive lenses. In this case, it is an assembly of concave lenses prepared by drilling aligned holes. However, both reflective and refractive optics work better at hard X-ray energies, above 5 keV. In the soft X-ray regime, the absorption of X-rays by elements increases drastically, thus, only diffractive optics can help modify the beam propagation.

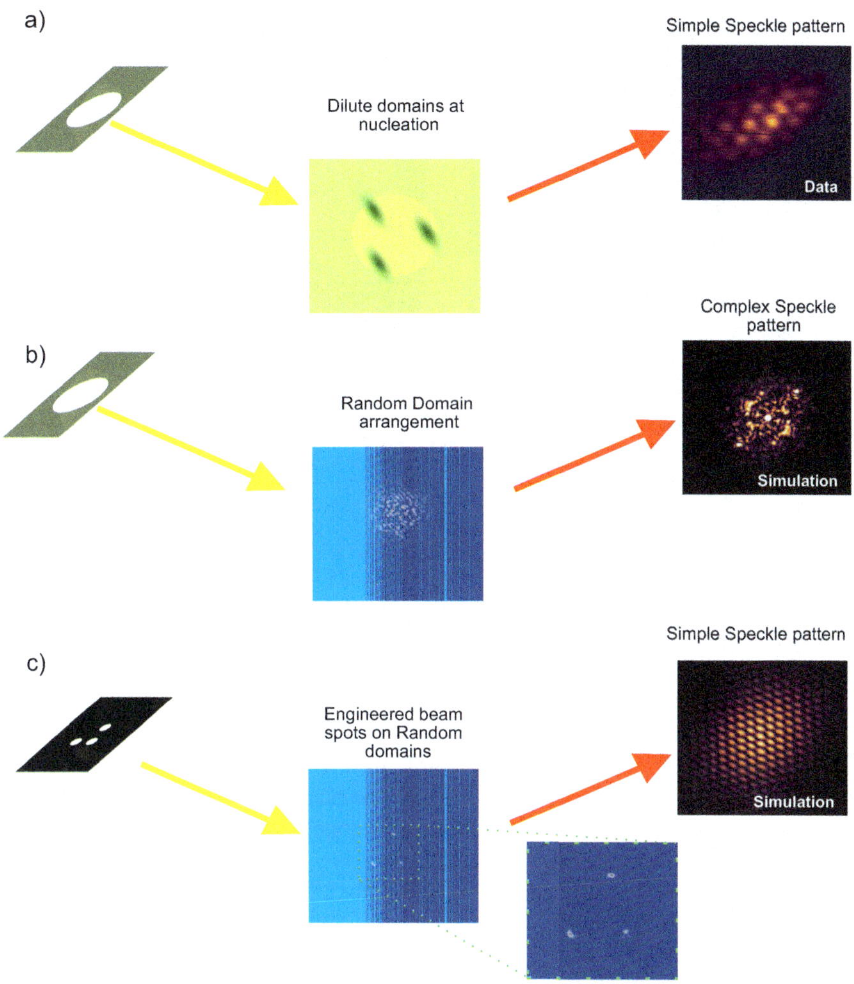

Fig. 5.1 Diluting the beam spot simplifies the speckle pattern:. a Shows our results of imaging dilute nucleating antiferromagnetic domains using pinhole scattering. **b** Demonstrates the increased complexity of the speckle pattern with more domains scattering under the beam-spot. The speckle pattern, an amplitude Fourier transform of the random domain arrangement, is simulated using [2] as a mask. **c** We propose diluting the beam spot into multiple, carefully arranged beam spots (three in this case) to restore the symmetry and simplicity of the speckle pattern

5.2.1 Fresnel Zone Plate

A Fresnel zone plate is a circular diffraction grating that focuses light similar to a lens [3]. The operating principle of a Zone plate is based on interference in the near-field limit. Before we discuss the zone plate technically, we discuss the design in simple terms. If light passes through a circular hole it produces annular interference fringes.

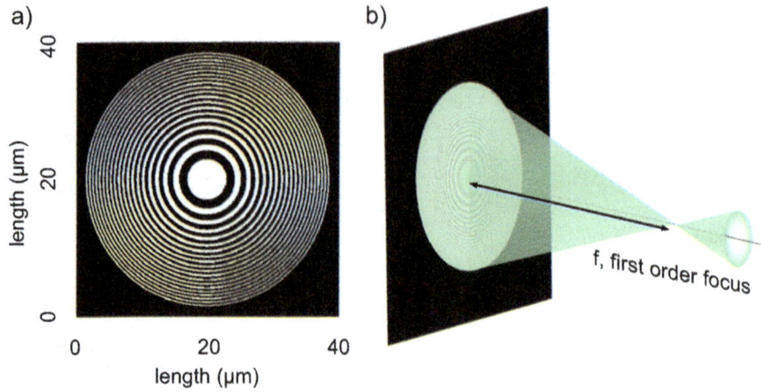

Fig. 5.2 Characteristics of a Fresnel zone plate:. a A typical zone plate. **b** Zone plate focuses the incoming beam into a tight focus spot. Here we have shown the beam propagation upto the first order focus and beyond

Now, if we back-propagate the wave from the rings of constructive interference, the phases must constructively match at the center of the hole. Thus, they will create a focus. Thus, if we illuminate a circular diffraction grating with a plane wave it is going to create focus by constructive interference. This is the basic idea which we will elaborate.

Fresnel zone plate acts in the near-field limit, where we must consider the curvature from the secondary wavefronts and cannot consider propagation from the sources as plane waves. Circular zones of a Fresnel zone plate are indexed by n up to N for the outermost zone. Let us consider that such a zone plate is illuminated by a plane wave of wavelength λ and has a focal length of f. Neighboring zones differ by a path length difference of $\lambda/2$. Let us also consider that n^{th} zone has a radius of r_n with a width of Δr_n. Then we have,

$$\sqrt{(r_n)^2 + f^2} - f = \frac{n\lambda}{2} \tag{5.1}$$

Expanding the square root and neglecting terms above second-order for a small zone plate we obtain (Fig. 5.2),

$$r_n^2 = n\lambda f \tag{5.2}$$

Here we have neglected the spherical aberration i.e. the fourth order term but this is a functional description of the zone plate. This is sufficient to see that the zone plate is inherently chromatic: every wavelength has a different focus. Thus, zone plate as a focusing optics is perfectly suited for coherent scattering chambers where high longitudinal coherence length requires filtering of wavelengths. The transverse coherence length, on the other hand, is achieved by focusing through constructive interference. The interference condition ensures phase matching of the incoming

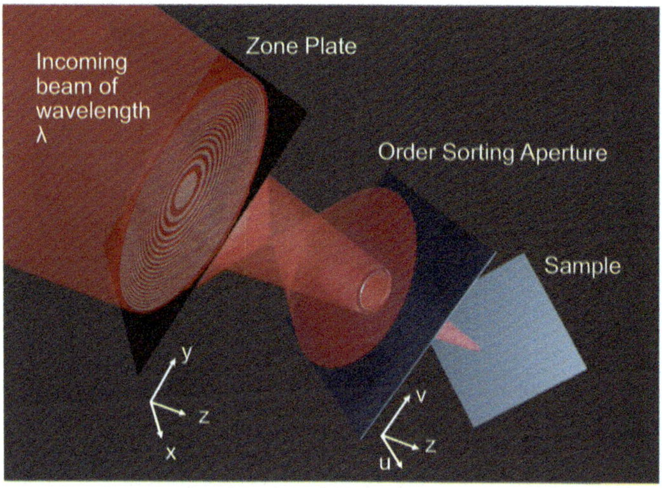

Fig. 5.3 Zone plate in experimental set up: One crucial instrument in zone plate-based coherent scattering is using order sorting aperture (OSA) which can be seen here between the zone plate and the sample. The order sorting aperture acts as a filter masking all other beam propagation-related different orders of focus but one. We see in the schematic drawing that the oSA blocks the beams corresponding to a higher order focus formed in between the zone plat and the OSA but a lower order focus passes through

wave fronts, and thus by construction they are coherent. Therefore, we focus on the zone plate as the element to work on for multiple beam-spot creation.

The above equation allows for the zone plate to have multiple foci along the principal axis of beam transmission. The mth order foci are to be found at $f/|m|$, with $n \rightarrow n|m|$, where $m \in \mathbb{Z}$. Thus along the axis zone plate creates cones of focused light propagating for each one of the focus and an unfocused beam for the 0th order. To have a single focused beamspot, we must filter out the other cones of different order. This is achieved by an order sorting aperture, which simply allows the desired order and blocks the other. Thus, a transmission Fresnel zone plate always comes with an order-sorting aperture. The full setup is shown in the Fig. 5.3.

Simulation of a zone plate wave propagation:

As discussed previously in Chap. 3 the propagation of a monochromatic coherent beam can be evaluated using scalar diffraction theory. Here we wish to simulate the beam profile propagation along the propagation axis for a given zone plate design illuminated by a coherent monochromatic beam. Rayleigh-Sommerfield equation as was discussed in the context of Eq. 2.9 will be used again and rewritten here as follows:

$$F(u, v) = \frac{\iota}{\lambda} \int_A f_{ZP}(x, y) \frac{e^{\iota \mathbf{k} \cdot \mathbf{r}}}{r} dx\, dy$$

Here $r = \sqrt{r_0^2 + (u-x)^2 + (v-y)^2}$ is the distance from a secondary source point in the zone plate (x, y) plane to a point in the sample plane or the observation plane (u, v). If we expand r in the limit where r_0 is considerably larger than the zone plate dimensions

$$F(u, v) = \frac{\iota}{\lambda} e^{\iota k r_0} \iint_{ZP} e^{-\iota \frac{\pi}{r_0 \lambda}(x^2+y^2)} e^{-\iota \frac{\pi}{r_0 \lambda} ux} e^{-\iota \frac{\pi}{r_0 \lambda} vy} f_{ZP}(x, y)\, dx\, dy \tag{5.3}$$

The term $e^{-\iota \frac{\pi}{r_0 \lambda}(x^2+y^2)}$ incorporates the curvature of the wavefronts into superposition from the propagated waves. In the far-field limit, where $(x^2 + y^2)/r_0 \ll \lambda$ that term approaches 1 and the far field distribution resembles a Fourier transform. The term $f(x, y)$ encodes the aperture function, which will be developed as the zone plates. Thus, using the Rayleigh-Sommerfield equation one can evaluate the propagated wavefront from the zone plates at any distance away from the zone plate surface.

Now we will set up the machinery to numerically forward propagate spherical wavefronts in transmission from the zone plate. For a given wavelength of plane wave illumination, we show below a zone-plate construction of 20 zones with $\Delta r_N \sim 200$ nm and an outer radius of 20 μm. We solve the Eq. 5.3 numerically to evaluate the beam propagation from an aperture function to an observation plane, which is our sample surface. To do this, we consider the impulse response of the aperture function. The impulse response is found by replacing the aperture function with a delta function in the origin. We denote the impulse response as $h(u, v) = \frac{\iota}{\lambda} \frac{e^{\iota k r}}{r}$. Hence, the equation for the super position of sources at sample plane

$$F(u, v) = \frac{\iota}{\lambda} \int_{Z} P f_{ZP}(x, y) \frac{e^{\iota k \sqrt{r_0^2 + (u-x)^2 + (v-y)^2}}}{r}\, dx\, dy$$

can be rewritten as a convolution integral

$$F(u, v) = \int_{Z} P f_{ZP}(x, y) h(u - x, v - y)\, dx\, dy \tag{5.4}$$

where ZP stands for doing the integration over the zone plate. Now if we take a Fourier transform on both sides and apply the convolution theorem of the Fourier transform we get -

$$\mathcal{F}(F(u, v)) = \mathcal{F}(f_{ZP}(u, v)) \mathcal{F}(h(u, v)) \tag{5.5}$$

Since convolution in real space is just multiplication in Fourier space Eq. 5.5 tells us that to evaluate the beam propagation, all we have to do is take the Fourier transform of the impulse response in the Fresnel limit, multiply that with the Fourier transform of the zone plate aperture function, and finally take an inverse Fourier transform to get F(u,v) at the sample plane. The impulse response now contains the information about the curvature of the wavefronts which are summed over the

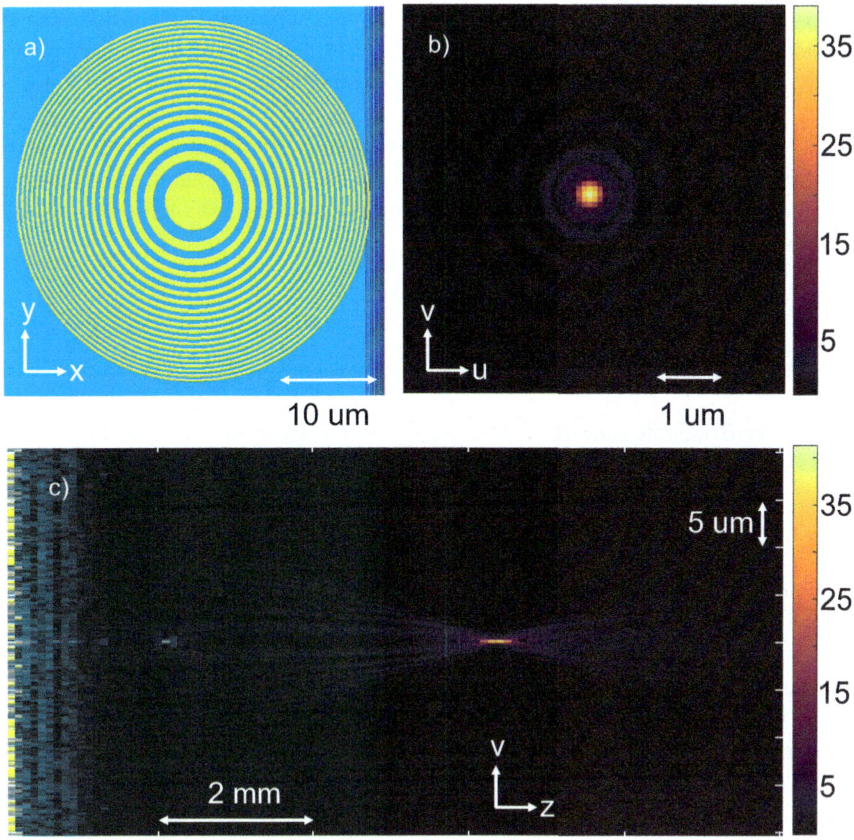

Fig. 5.4 Intensity propagation and profile at the focus for a typical Fresnel zone plate: a The real space design of the zone plate the bright regions are of pixel value 1 and the dark regions are of pixel value 0. The bright regions allow the incoming beam to pass while the dark region blocks them. We use this setup to create all aperture function zone plates here and discussed beyond. Each pixel of the 512×512 array represents linear dimension of 80 nm. **b** Intensity profile at the focal plane. The focal spot is approximately 300 nm while the outer ring radius was 200 nm. For conventional zone plates design the outer ring radius determines the focal spot size. **c** Intensity propagation along the wave vector of the beam propagation. The simulation shows the converging and diverging beams towards and away from two focal points: one, at 6.4 mm downstream, and the other higher order focus at around 2mm downstream of the zone plate. Each pixel roughly represents 100 μm. along z direction and 100nm in v-direction

superposition integral. Thus, we arrive at the equation that we use to find the beam propagation numerically (Fig. 5.4)

$$F(u, v) = \mathcal{F}^{-1}(\mathcal{F}(f_{ZP}(u, v))\mathcal{F}(h(u, v))) \quad (5.6)$$

In the simulation, we illuminate the zone plate with soft X-rays with energy 852 eV. We designed the zone plate on a 512×512 pixel frame with binary pixels. Each pixel has a dimension of 80 nm which is well above the fabrication limit of lithography nowadays. The focal length for the first-order focus is found at 6.4 mm. Also, we can see another higher-order focus around 2 mm. Furthermore, we show the intensity distribution at the focal plane where a clean single beam spot is visible. In nano-diffraction, transmission microscopy, and ptychography experiments such beam spot is utilized.

5.2.2 *Modified Fresnel Zone Plates*

Design-A: To deviate from the regular design and to create multiple beam spots, we argue on symmetry grounds. The symmetry of the function f(x,y) in general is also a symmetry of the function F(u,v). If we lower the symmetry of f(x,y) from C_∞ to C_2 the symmetry of F(u, v) will also be lowered. Based on this principle, we first show the design of a zone plate that can create two beam spots. The symmetry of the zone plate is lowered to create the two beam spots. One interesting outcome of this design is that the depth of focus is quite high. However, the efficiency is very low. Such a design may be useful for extremely bright coherent light sources such as 4th generation Synchrotron sources. However, we demonstrate proof of the principle on which we based our argument.

Increasing the number of beam spots is rather straightforward. To create three beam spots, we lower the symmetry to C_3. We have shown in Fig. 5.8 the construction of a three-beam spot focus from a modified zone plate design. The wavelength has always been 852 eV. Smallest pixel size is set to be 80 nm considering the lithography techniques used to fabricate the zone plates [reference]. However, advancement of the lithography techniques has helped to create high resolution and low defect zone plates. One of the experiments at Advanced Light Source achieved a resolution of 12 nm. The algorithm to create the zone plate designs starts from N holes symmetrically arranged around the focal point. Also, the arrangement of holes obeys the new symmetry we want it to obey. We then take the Fresnel transform and digitize the intensity distribution of the transform. This gives us the desired zone plate design. Taking a Fresnel transform of the aperture function gives us two beam spots in the soft X-ray regime. We note that this design is scalable easily to a higher number of beam spots. Figure 5.7 illustrates the algorithm to create the design responsible for creating multiple beam spots, as per design A. On the leftmost panel, we have shown two overlapped circular holes slightly off-centered from each other to break the axial symmetry. The boundaries of the circles are marked by red and blue dashed lines for illustration. First, we take a Fresnel transform of this aperture function to create the middle panel. The wavelength used for the transform is the same wavelength which we intend to use the zone plate with. The middle panel is what essentially gives us the desired aperture function which is responsible for the zone plate design. However, during the fabrication process using lithography, it is useful to have two well-defined

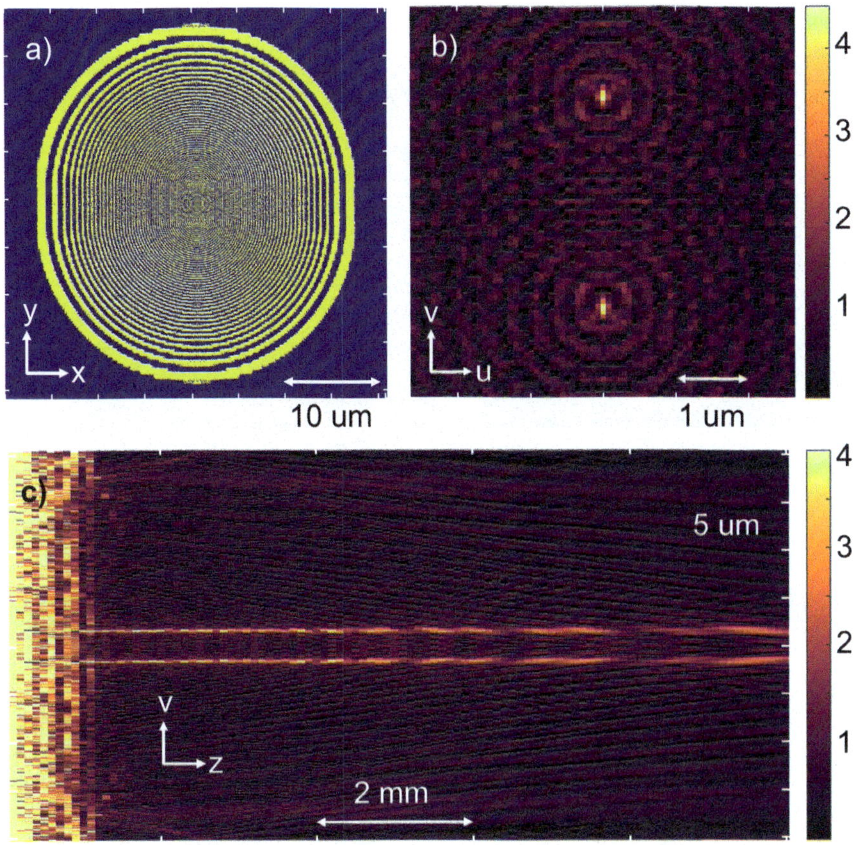

Fig. 5.5 Two Spot Creation: Design A **a** Design of the zone plate on a 512×512-pixel array with each pixel representing 80 nm. **b** Intensity profile observed 2 mm downstream of the zone plate where we can see two beam spots of 100 nm in size and are separated from each other by $\sim 3 \mu m$. **c** Beam profile along the wave vector of beam propagation showing a laser beam like beam creation

limits of transparency. We achieve that here by binarizing the middle panel above a threshold value of 0.58. The final design is shown on the right-most panel, which is again used in the demonstration Fig. 5.5 (Fig. 5.6).

Design B: Here we discuss another design to create multiple beam spots. This design is also scalable to a larger number of beam spots. Unlike previous argument, here we exploit the fact that if f(x,y) hosts a topological defect that is propagated in the wave field pass through the zone plate. Similar to our discussion in Chap. 3, where we observed that the topological defect of a spin texture leaves a fingerprint on the scattered beam, if we create a defect of winding number N, we observe $2 \times N$ lobes in the scattered beam. Angular symmetry of the defect on the zone plate design selects the correct Bessel mode, which translates the essence of the topological defect. Using this methodology, we can create $2N$ beam spots where $N \in \mathbb{Z}$. In the

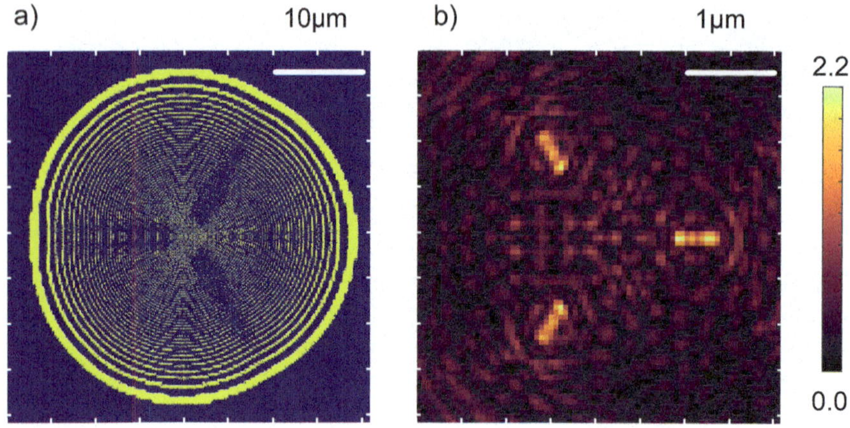

Fig. 5.6 Three Spot Creation: a The design of the zone plate (Design A) with each pixel representing 80 nm **b** beam profile on 3.7 mm downstream of the zone plate. Each spot is approximately 100 nm in size and comes in pairs. They are positioned on the vertices of an equilateral triangle of side length ~2μm. The OSA can be designed accordingly to select only one order

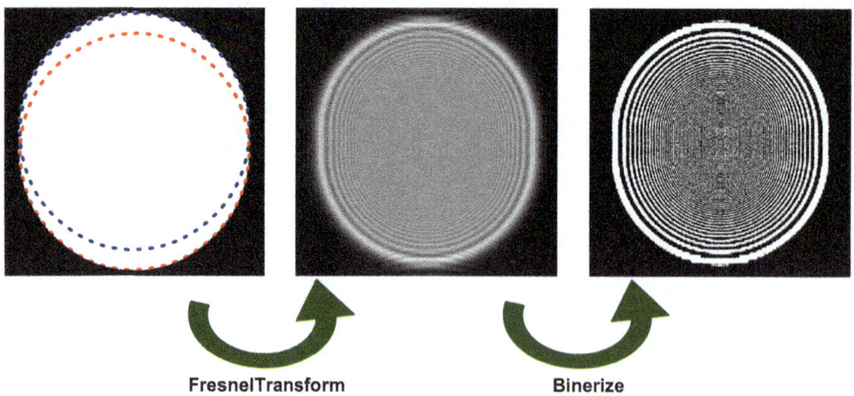

Fig. 5.7 Algorithm to create zone plates using design A

following illustrations, we show two such examples, where we have created 2, 4, and 6 beam spots. All of the designs discussed require a sorting aperture to limit higher-order contaminations. We denote that the efficiency of the zone plate decreased with number of beam spots increased.

To create the zone plate with a topological defect, we multiply the regular zone plate design with gradient masks as shown in the figure. The gradient masks can be defined by a function $\Theta(x - x_0, y - y_0)$, where (x_0, y_0) is the defect location. For our calculation, we have set

$$\Theta(x - x_0, y - y_0) = tanh(tan^{-1}\frac{y - y_0}{x - x_0}).$$

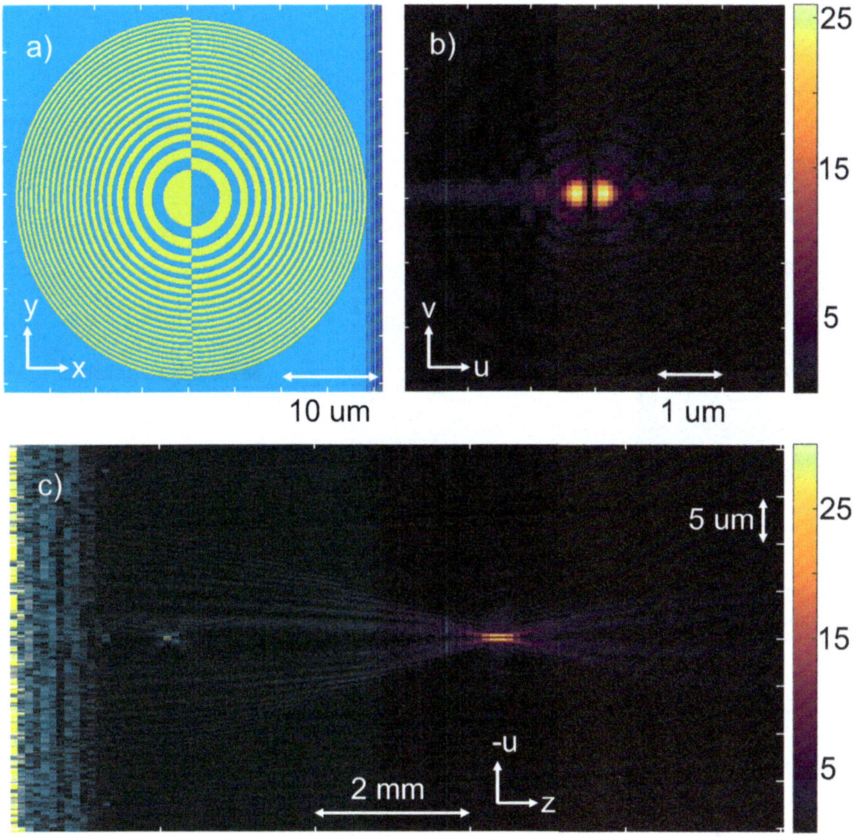

Fig. 5.8 Two Spot Creation: Design B a Design of the zone plate on a 512×512 pixel array with each pixel representing 80nm linear dimension. **b** The intensity distribution at the focal plane ~6 mm downstream. **c** Cross section of the space behind the zone-plate showing how focusing happens for this new design along the beam propagation (z)

Since $tanh(x)$ ranges between $(-1, 1)$, before multiplying with the zone plate design we rescaled them from $(0, 1)$. Then multiplying the zone plate with this function simply creates an envelope following the tanh function which has a singularity. This singularity is seen as the sharp contrast between the white and the black region in the middle panels. Since the gradient multiplied Zone plate blocks 50% of the beam we compensate that section with an inverted design of the same zone plate. Following the Babinet principle, we know that the diffraction pattern from the two inverted designs will be identical and will just add up. Each design contains the same defect, and thus the effect of the defect also survives. Finally, adding up, we get the design we have used to demonstrate the beam propagation along the axis in Fig. 5.8. The reason for creating a topological defect to split the beam can be understood from the point of view of Jacobi -Anger expansion. The angular symmetry of the zone plate selects

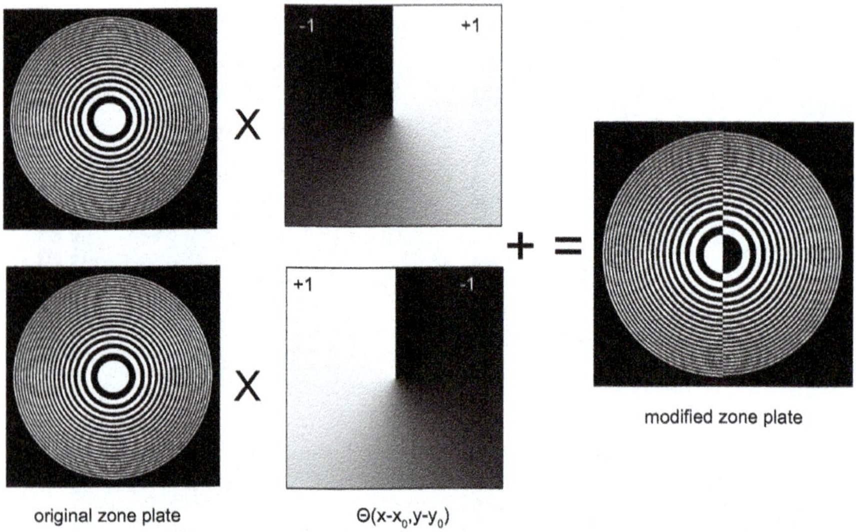

Fig. 5.9 **Algorithm to create the zone plate using Design B:** We take regular zone plates and impose a topological defect constraint. We sum two parts exactly opposite to each other to increase the output

the fundamental Bessel function to be propagated. The two-dimensional intensity distribution of Bessel functions of the first kind also obeys the same angular symmetry cast by the zone plate design. Any Bessel function other than the J_0 lacks intensity at the center and hence splits the intensity distribution. Thus, from an isotropic intensity distribution at the focal plane, we get a beam spot with two split beam spot (Figs. 5.9 and 5.10).

Thus, we have shown that coherent beam spots can be created with modified zone plate constructions. To design these modified zone plates we argued based on the symmetry and topology of the aperture function and its influence on the propagated beam. There are simpler ways in which one can create two beamspots too. For example, just by blocking off the beam using two slits or two pinholes one can in-principle create multiple spots. Although, if the slits are tightly spaced, higher order diffraction effects will show up on top of the fundamental aperture function. Since we already utilize diffraction effect in designing the zone plates, the designs are free from such aberrations.

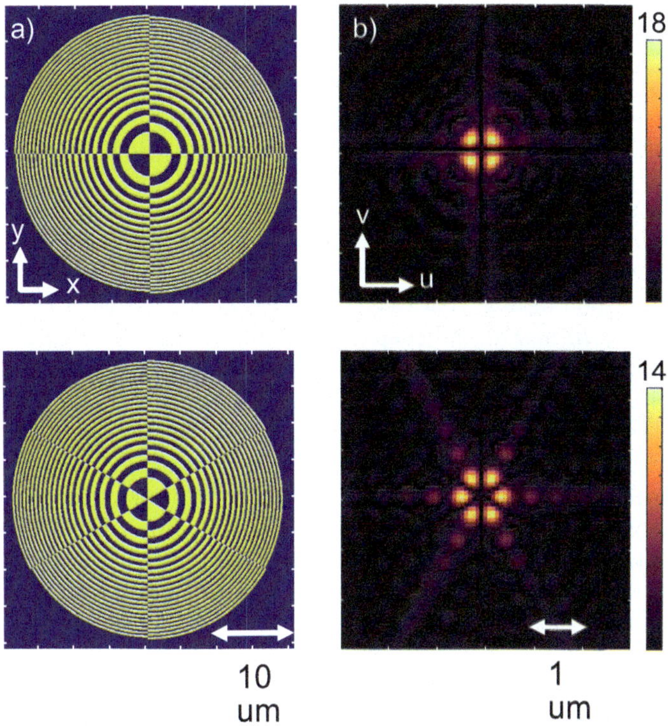

Fig. 5.10 Four and six Spots Creation using Design B a the zone plate design scaled to produce 4 and 6 beam spots at the vertices of a symmetrical polygon. **b** Intensity distribution at the focal plane

5.3 Discussion: Using Multiple Coherent Beam Spots

5.3.1 Enhancement of Existing Experiments

Now that we have created two beamspots we move on to discuss the benefit of utilizing multiple beam spots in our experiments. The fundamental advantage of using two beam spots over a single beam spot comes from the simplicity of the scattered speckle pattern. As we saw previously, the speckle pattern arising from such dilute domain distribution helps us to extract the distribution of domains without a phase reconstruction algorithm. Thus, we expect that a dilute distribution of coherent beam footprint will simplify the existing experiments we carry out on our samples. To illustrate this point, we show here two experiments where we take advantage of the two beam-spot coherent scattering.

Domain Orientation

It is often found that the mesoscopic domain structures have a preferential direction for growth after nucleation [2, 4]. For example, as shown in the case for $NdNiO_3$ thin film measured with X-PEEM. The step terraces of the substrate provided the motif in this case, and thus 200nm thick but long structures of thin films were deposited. Individual strips eventually coalesce together, but provide a growth barrier for the nucleating insulating domains below the metal-insulator transition as shown here in Fig. 5.11. Now, X-PEEM is a microscopy technique so we could see the preferential domain orientation readily. However, if we were to do a coherent resonant X-ray scattering we will tune to the coupled antiferromagnetic order emerging in the insulating states and let us assume that they would follow the same mesoscopic distribution as the insulating domains. Then deciphering the domain orientation will not be easy from the scattered data. In the figure, we simulate the single spot scattering experiment. With many domains participating in the scattering process, we see a complex speckle pattern. We observe two major high intensities in the speckle pattern and they are arranged with two different symmetries. To understand the relative role of the correlations we need to characterize further possibly with a four-point correlation resolved in angle space. However, we show below that with two beam spot experiments, this observation becomes readily interpretable from the scattered

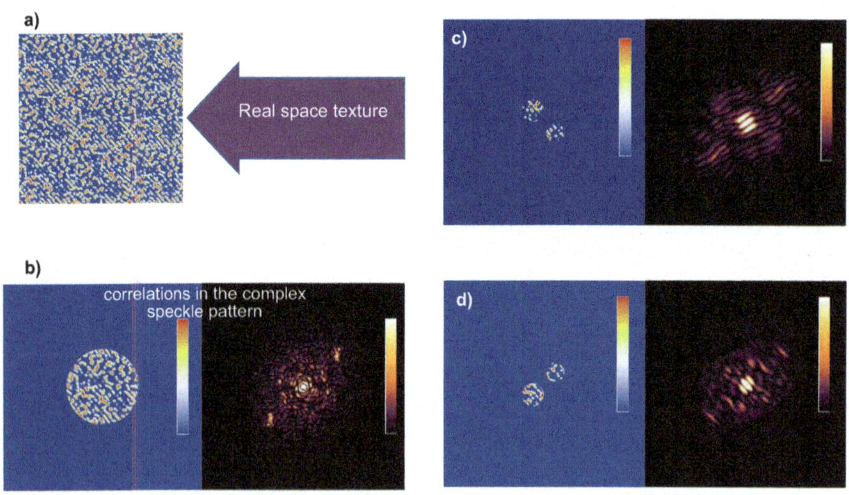

Fig. 5.11 Demonstration through Simulation I a a real space domain texture created from using experimental phase heterogeneity observed by Mattoni et al. during an X-PEEM experiment[2]. The preferential orientation of the domains are along the diagonal of the panel **b** Simulation of coherent scattering outcome if we do single beam-spot coherent scattering from this texture. The result is a complex scattering pattern **c–d** Two beam spot scattering from the domain patterns with two different orientations of the beam spots relative to domain orientations. When the separation vector between the beam spots is parallel to the modulation direction we see clear supermodulation peaks in the speckle pattern as seen in **d**

intensity. In Fig. 5.12 we discuss scattering from the two beam spots as we rotate the sample. With the preferential domain growth direction aligned perpendicular to the separation vector of the beam spots, we can clearly see the high intensity as the dominant super-modulation on the peak. On the other hand, if the beam spots are arranged parallel to the domain anisotropy it hides the information under the complex correlation.

Thus, we have shown here that we can use the two beam spots instead of one as a spatial filtering mechanism to simplify the experiment at hand. Although we argued for a two-fold (C_2) symmetry of domain orientation, we can easily scale the experiment for a different C_n symmetry of domain orientation.

Single Shot Imaging

To study the mesoscopic organization and evolution single shot imaging is essential. We have established this over and over throughout the development of this thesis. However, depending on the contrast we saw that the single shot speckle pattern can be either studied with a phase reconstruction algorithm—that suffers from multiple possible solutions or, resort to a higher order correlation of the speckle pattern. However, holography gets around the problem of phase information lost by the use of a reference beam. With the two beam spots, we show here that we can simulate a similar scenario. The advantage is of course the simplicity in finding the real space image. In holography, one usually drills a fine enough hole through the sample to have a reference beam interfere with the transmitted beam scattered from the heterogeneous sample. One simply takes a Fourier transform of the speckle pattern observed on the detector to get the real space image.

We have shown such a scenario with two beam spots where one beam spot illuminates a region with a complex domain arrangement, while the other illuminates only a single domain. Such an experiment can be designed first by knowing the correlation length of the domains at the temperature of the experiment from incoherent scattering data, and then following it up with the construction of an asymmetrical two-beam spot experiment. For the case of devices where one is certain of the interested region, the problem of creating a single domain under one of the beam spots can be solved by utilizing focused ion beam damage as well. We can see from our simulated illustration that the image of the domain arrangement could be easily found from the inverse Fourier transform. Since the scattered signal is convoluted with the reference domain, we estimate a resolution on the order of the reference domain size.

Such an experiment is possible with single-spot experiments, but we are often bound to image the location close to the reference spot which has suffered from fabrication techniques such as focused ion beam milling. For strongly correlated electronic systems, such perturbations may change significantly the electronic character of the area under investigation. With two beam-spot constructions, we can separate the beam spots far enough that the fabricated area does not interfere with the area imaged.

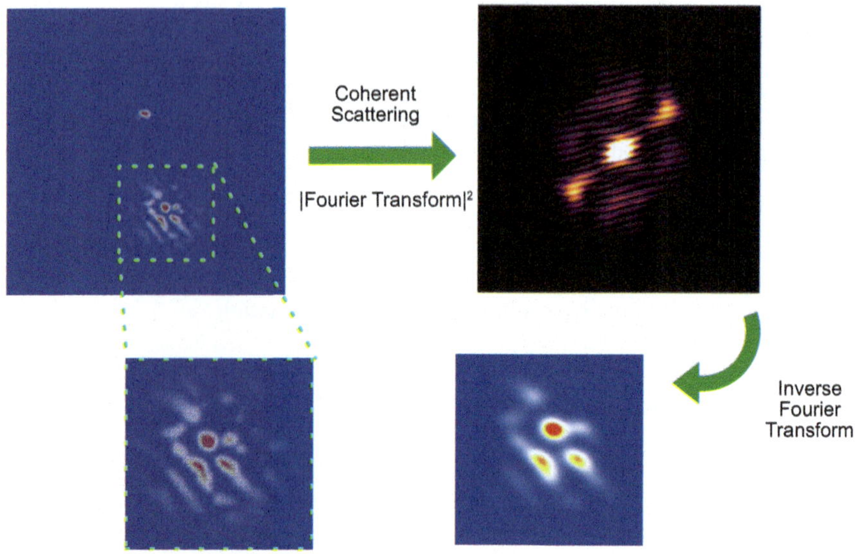

Fig. 5.12 Demonstration through simulation II—Single shot imaging using two beam spot coherent scattering: a We show the real space texture contributing to the speckle pattern after two beam spots illuminate the domain distribution in the sample. If under one beamspot there is only one domain then the experiment essentially becomes a holography experiment convoluted with single-domain scattering instead of a reference beam. Thus the image of the scattering complex domain distribution can be found simply by taking the inverse Fourier transform on the observed speckle pattern (**b**), evaluated here by an amplitude Fourier transform. The resolution of the imaging experiment will be limited by the single domain size under the top beam-spot **c** zoom in on the real space complex domain arrangement **d** inverse Fourier transform of the speckle pattern. The similarity between **c**–**d** proves the imaging aspect of this experiment. The texture is adapted from [2]

Lateral Structure of Emission

In 4th generation Synchrotron sources, the coherence is significantly high. It has been recently shown that such an incoming beam with a long longitudinal coherence length (3 μm) can image the fluorescence centers [5–7]. To show this, the authors created two beam spots on a Cu foil of 20 μm thickness. The two beam spots were created using compound refractive lenses along with phase grating to split the beam. The incoming pulses of 10 fs duration and photon energy of 9 keV pass in transmission through the Cu foil before it is recorded in an area detector. We have reprinted their experimental details in Fig. 5.13 to facilitate the discussion. The photon energy was chosen to be just slightly above the Cu $K\alpha_1$ and $K\alpha_2$ to facilitate fluorescence emission. They found that the intensity-intensity correlation, $g^2(q)$ obtained from each one of the 58 million recorded frames showed the fringe pattern expected from two fluorescing centers, just like double-slit interference.

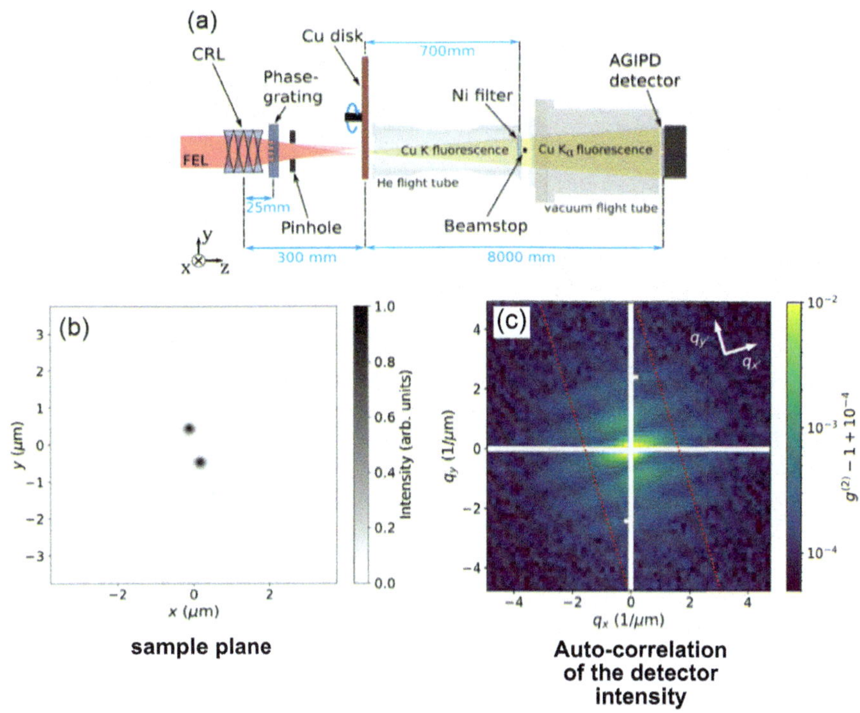

Fig. 5.13 **Imaging via correlation of X-ray fluorescence** [5]: **a** The experimental set-up splits the coherent beam using CRL and phase grating. The expected beam spot profile is shown in panel **b** where we see two beam spots separated by about a micron. The FEL pulses pass through the 30 μm Cu-foil and the foil fluoresce at Cu K α edge shown in **a**. The detector collected 58 million frames of intensity patterns with very low contrast. However, taking auto-correlation on the pattern the authors could find fringe patterns related to the fluorescence centers. Thus from the auto-correlation patterns, they could reconstruct the image of the two spots

The lifetime of the core-hole excitation associated with the fluorescence emission is 0.6 fs. They were generated over the time of 67 fs which is required for the incoming pulse to travel the thickness of the Cu foil. However, this was only possible because of the high brilliance, compensating for the low signal-to-noise ratio. Another way to compensate is probably creating multiple beam spots in an array instead of two beam spots, which we would like to discuss here.

That beam spot engineering in coherent sources can extract information previously unseen, has been comprehensively demonstrated by the experiment discussed [5]. If the spectral line width is small on the order of a few meV one can potentially get a coherence time on the order of 1ns, which brings us to the realm of the fastest area detector available for 3rd generation Synchrotron sources. Utilizing these detectors in third-generation Synchrotron sources, we propose to create an array of beam spots with well-defined separation on the sample. This is, ideally, a grating on the sample surface can resolve energies. Now, we set an incoming beam to one of the resonance

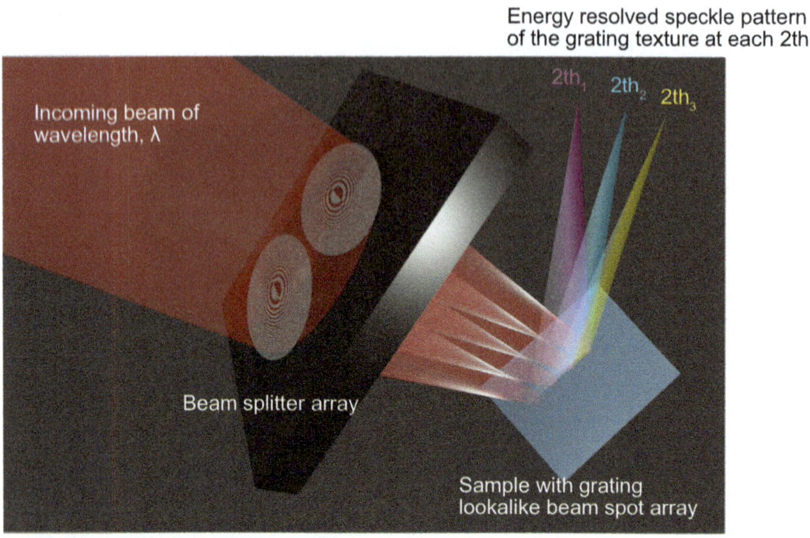

Fig. 5.14 Combining resonance inelastic X-ray scattering with coherence: An array of two beam-spot creating zone plate can create an array of equally spaced beam spots on the sample. This grating structure will resolve scattered intensity by energy in reciprocal space at different 2θ locations. Thus the coherent scattering pattern observed will be energy resolved

edges of the sample. This not only increases the photon count on the output but also creates a core hole near the valence shell of the electron which relaxes back following all possible electronic excitation pathways available in the system. Let us suppose for the sake of discussion that we have a strongly correlated electronic system with a transition metal ion of Ni. If we create a $2p \rightarrow 3d$ transition using Ni-L3 edge, it will emit at the energy of Ni-L3 along with Ni-L3+1 eV for the dd excitation, and all other possible emission lines. Those lines, that have a very sharp bandwidth on the order of 1–10meV, can be studied. Under the grating, the emission lines will create their diffraction peaks at different angular positions in space. Thus, to select a specific energy we need to park our detector at the respective 2θ angle (Fig. 5.14).

The first-order or second-order correlation in the intensity pattern recorded will be able to tell us the real space distribution of the emission scattering centers. Furthermore, we can correlate this real space information with the elastic line which will be resolved similarly. The elastic contribution will tell us information similar to energy-integrated coherent scattering. Thus, in this way, we propose to resolve the speckle pattern in the energy space.

5.3.2 New Mode of Coherent Experiments for Driven Systems

Here we explore the possibility of utilizing two-spot experiments as a possible probe for studying systems under dynamic equilibrium. Even though we had been studying the static phase organization of the samples, in practice condensed matter systems are often used for the voltage and current response from the material. Thus, understanding physical systems under the gradient of potential is of considerable importance. In the context of temperature one can think of a situation where the base temperature of the sample is set at T while the left edge is connected to a source with T+ΔT and the right edge is connected to a sink of T-ΔT. Such a system will be coupled to a thermal current. We can subsequently design experiments to probe the evolution of the domain organization to study the coupling. With the directionality in-built in the two beam-spot we can easily employ such tools to study these problems.

Examples of Directionally Driven Systems

The design of magnetic racetrack memory is built around this idea of a directionally driven system [8]. The authors showed that short pulses of spin-polarized current can be used to drive the domain walls in a permalloy nano-wire. The domain arrangement of alternate magnetic moments can be used to load non-volatile memory into the nano-wire. Then the domain arrangement can be driven from one end while the data is read out at a different suitable position along the 1-dimensional distribution of domain arrangement. This is precisely the kind of system we wish to study. Another similar work has been shown in the context of topological defects, such as skyrmions in a 1-dimensional channel of metallic ferromagnets [9]. Here, the current pulses coupled to the skyrmions and moved them at speeds exceeding 100 ms^{-1}. Both of these experiments will show more complexity in higher dimensional version where other degrees of freedom will be important and will influence the kinetics significantly. If we consider that only energetics drive the equation of motion for the domain walls and skyrmions then not only observing their kinetics will tell us about the fabrication quality of the next-generation magnetic storage system, but will provide us with a direct experimental methodology to study the driven system in a dynamic equilibrium. On top of the energy landscape driving most of these kinetics, all motion will suffer from fluctuations and dissipation which will make the kinetics richer in phenomenology.

To understand the kinetics, we need to devise an experiment that can be done in a single shot. Then the time evolution can be recorded with subsequent shots. Coherent scattering is such an experiment that collects data about the phase organization in a single shot. However, since the data is collected over a large region it suffers from complexity as discussed above. Thus, we propose to simplify the result using multiple beam spots where the beam spots are arranged according to the science question one is looking for.

Illustration of the New Mode Schematically

A demonstration of a hypothetical experiment is shown below using two-beam spot coherent scattering where we can see the scattering output change due to the domain motion as shown by an arrow in the second plot. Here we consider the motion of the domains towards a direction noted by the arrows. Left and right columns correspond to a former and later instants of time, respectively. For each instant of time, we show the real space domain distribution and the Fourier transform depicting the coherent scattering outcome. We vary the beam spot shape between the top and bottom rows.

On the top row of the panels, we show the speckle pattern that will be observed had we experimented with a coherent beam spot similar to the circle drawn on each real space panel. The left and right panels are two instants of real space texture and corresponding coherent scattering outcome. The two reciprocal space speckle patterns can be considered two shots of the experiment. We note that the speckle pattern remains relatively similar. However, the distinct scattering outcome can be observed from the bottom panels, where two smaller beam spots were used to filter out all other information. The fringe pattern for the bottom-right panel in the reciprocal space tells us about the separation of the domains. By tracking the fringe pattern on the detector, thus, we can record the kinetics of the domain motion.

A single large beam-spot coherent scattering gets all possible interference patterns that can be generated from individual domains. Thus, the coherent scattering results on the top row are complex and hardly reveal any information from the single-shot mode about the motion the domains underwent. The power of diluting the beam spot comes from spatially filtering the scattering field to simplify the scattered speckle pattern. Designing two nano-beam spots helps us create a well-separated and well-directed scattering filter from the sample. Simplification makes it possible to directly observe the evolution from single shot coherent scattering outcome (Fig. 5.15).

5.4 Conclusion

In this chapter we have shown that if we create multiple beam spots of coherent beam, we can expand our experimental methodology to study domain dynamics at mesoscale. Furthermore, these multiple beam spot creation help us simplify existing experiments for a direct single shot observation of the experimental question at hand. Thus, it can be easily expanded to a time domain characterization.

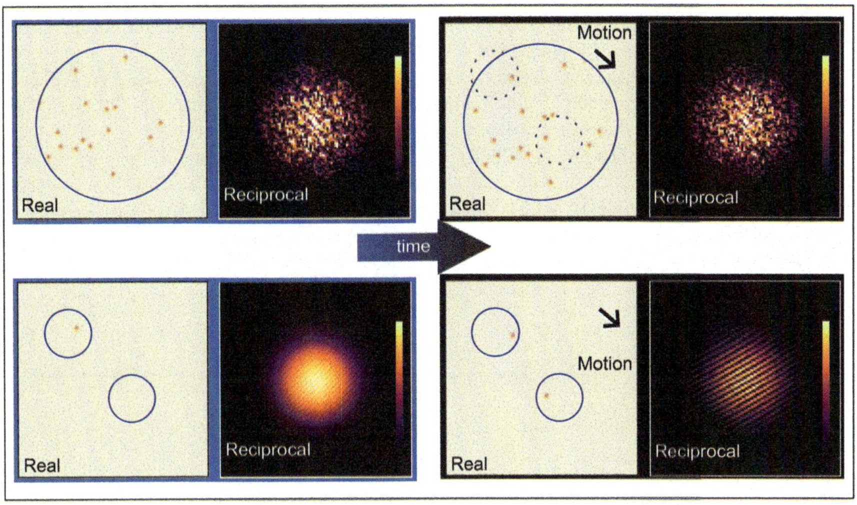

Fig. 5.15 Demonstration through simulations III—New mode of coherent scattering experiment where we can study the kinetics of real space textures:. The top row shows two instances of the regular coherent scattering experiment using a single beam spot created with a pinhole. The real space domain distribution moves along the directed motion arrow however the complexity of the speckle pattern fails to capture it. The same domain motion is now studied using two-beam spot scattering. The single-shot coherent scattering outcome considerably changes indicating a phase texture change. The real space figure in the top right panel shows the relative comparison between the beam spot structures where the two-beam spot is shown with dashed lines

References

1. Bluschke M, Basak R, Barbour A, Warner AN, Fürsich K, Wilkins S, Roy S, Lee J, Christiani G, Logvenov G, Minola M, Keimer B, Mazzoli C, Benckiser E, Frano A (2022) Imaging mesoscopic antiferromagnetic spin textures in the dilute limit from single-geometry resonant coherent x-ray diffraction. Sci Adv 8(29):eabn6882. American Association for the Advancement of Science
2. Mattoni G, Zubko P, Maccherozzi F, van der Torren AJH, Boltje AB, Hadjimichael M, Manca N, Catalano S, Gibert M, Liu Y, Aarts J, Triscone J-M, Dhesi SS, Caviglia AD (2016) Striped nanoscale phase separation at the metal–insulator transition of heteroepitaxial nickelates. Nat Commun 7(1):13141. Nature Publishing Group
3. Jacobsen C, Howells M, Warwick T (2019) Zone-plate x-ray microscopy. In: Hawkes PW, Spence JCH (eds) Springer handbook of microscopy. Springer International Publishing, Cham, pp 1145–1204
4. Wu L, Shen Y, Barbour AM, Wang W, Prabhakaran W, Boothroyd AT, Mazzoli C, Tranquada JM, Dean MPM, Robinson IK (2021) Real space imaging of spin stripe domain fluctuations in a complex oxide. Phys Rev Lett 127:275301
5. Trost F, Ayyer K, Prasciolu M, Fleckenstein H, Barthelmess M, Yefanov O, Dresselhaus JL, Li C, Bajt S, Carnis J, Wollweber T, Mall A, Shen Z, Zhuang Y, Richter S, Karl S, Cardoch S, Kumar Patra K, Möller J, Zozulya A, Shayduk R, Lu W, Brauße F, Friedrich B, Boesenberg U, Petrov I, Tomin S, Guetg M, Madsen A, Timneanu N, Caleman C, Röhlsberger R, von Zanthier J, Chapman HN (2023) Imaging via correlation of x-ray fluorescence photons. Phys Rev Lett 130:173201

6. Trost F, Ayyer K, Chapman HN (2020) Photon statistics and signal to noise ratio for incoherent diffraction imaging. New J Phys 22(8):083070. IOP Publishing
7. Trost F, Ayyer K, Oberthuer D, Yefanov O, Bajt S, Caleman C, Weimer A, Feld A, Weller H, Boutet S, Koglin J, Timneanu N, von Zanthier J, Röhlsberger R, Chapman HN (2023) Speckle contrast of interfering fluorescence X-rays. J Synchrotron Radiat 30(1):11–23
8. Stuart SP Parkin MH, Thomas L (2008). Magnetic domain-wall racetrack memory. Science 320(5873):190–194
9. Woo S, Litzius K, Krüger B, Im M-Y, Caretta L, Richter K, Mann M, Krone A, Reeve RM, Weigand M et al (2016) Observation of room-temperature magnetic skyrmions and their current-driven dynamics in ultrathin metallic ferromagnets. Nat Mater 15(5):501–506

Appendix A
Simulating Dilute Domain Distribution and Textures

```
"THIS CODE SHOWS COHERENT DIFFRACTION OUTCOME OF DOMAIN ARRANGEMENT OF underlying FM/AFM ORDER"
FOV=2;
(*Toggle from Gaussian circular FOV\[Equal]2, Full FOV\[Equal]1 , to chosen nanospot(s) FOV\[Equal]0]*)
noOrder=0;
(*Toggle it to 1 if you don't want any order in the lattice - chooses random U(1) member for every site*)

DefMany=0;
(*If the lattice has many Topological defects, DefMany=1, else DefMany=0*)
winding=0;
(*For a central winding choose DefMany=0*)
topRight=0;botRight=0;botLeft=0;
(*For phasefaults, DefMany=0, winding=0*)

Defect=0;
(*Turn it 1 if you want randomly placed annihilition on a lattice*)
Jitter=0; Angl=1*\[Pi]/18;
(*For random vector added to the design to change the correlation length*)

size=420;
(* Enter Lattice Size*)
LatticeU=ConstantArray[0,{size,size}]; (*Create the Blank Lattice*)
Lattice=ConstantArray[0,{size,size}]; (*Create the Blank Lattice*)
c=2;
```


R. Basak, *Nucleation, Fabrication, and Competition in Antiferromagnetic Textures Studied with Coherent Resonant Soft X-ray Scattering*, Springer Theses,
https://doi.org/10.1007/978-3-032-19199-1

```
(*For Ferromagnetic texture use c = 1, for Antiferromagnetic c = 2*)

If[noOrder==1,

For[i=1,i<= size,i++,
For[j=1,j<= size,j++,
LatticeU[[i,j]]=Exp[-I*(2*\[Pi])/1*RandomReal[]]]]; (* Defined the underlying lattice LatticeU*)
,Print["There is an order"]];

If[noOrder==0,
(*Declare the wavelength of lattice order. For FM this is 1, for AFM 2xInteger*)
For[i=1,i<= size,i++,
For[j=1,j<= size,j++,
LatticeU[[i,j]]=Exp[-I*(2*\[Pi])/(1*c)*Mod[(i-size/2)+1*(j-size/2),1*c]]]]; (* Defined the underlying lattice LatticeU*)
,Print["There is no order"]];

(*Change 0 \[Rule] 1  in front of (j-size/2) for a different k here and a different k below*)
(*Change -1 \[Rule] 1 in front of \[ImaginaryI]*(2*\[Pi])/(1*c) for a different cycloid between domains below*)
(*Change -1 \[Rule] 1 in Global Phase to make it Antivortex for a different cycloid between domains below*)


(* Note: The following lattice introduces line defect - this is phase fault// Top Right *)
If[topRight==1,
For[i=size/2+1,i<= size,i++,
For[j=size/2+1,j<= size,j++, (* by changing this j limit(1 \[Rule] size/2) one can make the phases share (1\[Rule] 2) edge*)
LatticeU[[i,j]]=1/1*Exp[I*(2*\[Pi])/2]*Exp[(-1)*I*(2*\[Pi])/(1*c)*Mod[(i-size/2)+1*(j-size/2),1*c]]]],Print["No TOp-Right phase fault"]]; (* Defined the underlying lattice LatticeU*)
(* Note: The lattice has line defect build into it - this is phase fault*)

(* Note: The following lattice introduces line defect  - this is phase fault// Bottom Right*)
If[botRight==1,
For[i=size/2+1,i<= size,i++,
For[j=1,j<= size/2,j++, (* by changing this j limit (1 \[Rule] size/2) one can make the phases share (1\[Rule] 2) edge*)
```

```
LatticeU[[i,j]]=1/1*Exp[I*(2*\[Pi])/2]*Exp[(-1)*I
    *(2*\[Pi])/(1*c)*Mod[(i-size/2)+1*(j-size/2),1*c
    ]]]],Print["No Bottom-Right phase fault"]]; (*
    Defined the underlying lattice LatticeU*)
(* Note: The lattice has line defect build into it -
     this is phase fault*)

(* Note:The following lattice introduces line defect
     - this is phase fault// Bottom Left*)
If[botLeft==1,
For[i=1,i<= size/2,i++,
For[j=1,j<= size/2,j++, (* by changing this j limit
    (1 \[Rule] size/2) one can make the phases share
    (1\[Rule] 2) edge*)
LatticeU[[i,j]]=Exp[I*(2*\[Pi])/1]*Exp[(-1)*I*(2*\[
    Pi])/(1*c)*Mod[(i-size/2)+1*(j-size/2),1*c]]]],
    Print["No Bot-Left phase fault"]]; (* Defined the
     underlying lattice LatticeU*)
(* Note: The lattice has line defect build into it -
     this is phase fault*)

(*
LatticeU[[128,128]]=Exp[-\[ImaginaryI]*\[Pi]/2]*
    LatticeU[[128,128]];
LatticeU[[127,128]]=Exp[-\[ImaginaryI]*\[Pi]/2]*
    LatticeU[[127,128]];
LatticeU[[129,128]]=Exp[-\[ImaginaryI]*\[Pi]/2]*
    LatticeU[[129,128]];
LatticeU[[128,127]]=Exp[-\[ImaginaryI]*\[Pi]/2]*
    LatticeU[[128,127]];
LatticeU[[128,129]]=Exp[-\[ImaginaryI]*\[Pi]/2]*
    LatticeU[[128,129]];*)


(* Writing Defects: 2A: This is a continuous
    deformation of the magnetic texture*)

If[DefMany==0 && (topRight+botRight+botLeft)==0 &&
    noOrder==0,
For[i=1,i<= size,i++,
For[j=1,j<= size,j++,
LatticeU[[i,j]]=Exp[I* winding*ArcTan[(i-size
    /2.00002),(j-size/2.00002)]]*Exp[-I*(2*\[Pi])/(1*
    c)*Mod[(i-size/2)+1*(j-size/2),1*c]]]]
,Print["Working on multiple spots"]];
(* Defined the underlying lattice LatticeU*)


"Underlying Lattice:"
ListVectorPlot[Table[{{m,n},{Re[LatticeU[[m,n]]],Im[
    LatticeU[[m,n]]]}},{m,Round[size/2]-2*c,Round[
    size/2]+2*c,1},{n,Round[size/2]-2*c,Round[size
```

```
    /2]+2*c,1}],VectorPoints->All,VectorMarkers->"
    CircleArrow",VectorScale->0.05]

(*
ListVectorPlot[Table[{{m,n},{Re[LatticeU[[m,n]]],Im[
    LatticeU[[m,n]]]}},{m,164-2*c,164+2*c,1},{n
    ,104-2*c,104+2*c,1}],VectorPoints\[Rule]All,
    VectorMarkers\[Rule]"CircleArrow",VectorScale\[
    Rule]0.05]
*)

(*__Writing a Topological Defect__*)


(*
HOW TO USE THE CODE:Entry 2: DESIGNING THE DOMAIN
    CLUSTERS: DEFINE A RELATIVE POSITION
Comment 4,5 and 6 to see what happens when there is
    triangular arrangement instead of hexagonal
    arrangement.
 Here and in HOW TO USE THE CODE: Entry 3 - -
     Originally 1-2-3 corresponds to 3:7:11 o'clock
     location and 4-5-6 corresponds to 9:1:5 o'clock
     location.
After 35 deg cw rot 1-2-3 corresponds to 2:6:10 o'
    clock location and 4-5-6 corresponds to 8:12:4 o'
    clock location.
*)

(*mask=0*IdentityMatrix[size]; (*Creating the mask
    function makes it possible to define a domain
    arrangement of the order defined in LatticeU*)

mask[[100;;120,100;;120]]=1;
mask[[size-120;;size-100,size-120;;size-100]]=1;
Image[mask]*)
(*Grid[{{Image[Re[Lattice],ImageSize\[Rule]Medium],
    Image[Im[Lattice],ImageSize\[Rule]Medium],
    MatrixPlot[Lattice,ImageSize\[Rule]Medium]}},
    Frame \[Rule]All]
"Left Image: shows locations of 1, Middle Image:
    shows location of i and Right Image: locations of
     +1 and -1"
*)

(*Change sc=3 for Full FOV=1*)
sc=2; LatCoLen=21;
\[CapitalDelta]=21*sc;

    (*Creating 6 location points corresponding to a
    hexagonal arrangement of domains, \[CapitalDelta]
     is side length. For triangle \[CapitalDelta]\[
    Sqrt]3 is the side length*)
```

```

PtDef1x=size/2;PtDef1y=size/2+\[CapitalDelta];
PtDef2x=size/2+(\[Sqrt]3)(PtDef1y-PtDef1x)/2;
PtDef2y=PtDef1x-(PtDef1y-PtDef1x)/2;
PtDef3x=size/2-(\[Sqrt]3)(PtDef1y-PtDef1x)/2;
PtDef3y=PtDef1x-(PtDef1y-PtDef1x)/2;
(*PtDef4x=size/2;PtDef4y=size/2-\[CapitalDelta];
PtDef5x=size/2-(\[Sqrt]3)(PtDef1y-PtDef1x)/2;
PtDef5y=PtDef1x+(PtDef1y-PtDef1x)/2;
PtDef6x=size/2+(\[Sqrt]3)(PtDef1y-PtDef1x)/2;
PtDef6y=PtDef1x+(PtDef1y-PtDef1x)/2;
*)

(*
HOW TO USE THE CODE:Entry 3:DESIGNING THE DOMAIN CLUSTERS: DEVIATE FROM DEFINED POSITIONS BY GLOBAL ROTATION, INDIVIDUAL ROTATIONS and SEPARATIONS
Comment 4,5 and 6 to see what happens when there is triangular arrangement instead of hexagonal arrangement.
\[Theta] is global rotation to all domains, except the one lying at the center. \[Phi]1 and \[Phi]2 generate individual rotations Rot1 and Rot2 which can be given to
a pair of anti-positioned domains/defects. \[Delta] is the individual unit displacement which can be added to the locations at will.

*)

\[Theta]=(5*\[Pi]/180);(*Global Rotation of domains *)
\[Phi]1=(0*\[Pi]/180);(*Rotation of a domain wrt others*)
\[Phi]2=(0*\[Pi]/180);(*Rotation of another domain wrt others*)
\[Delta]=0; (* Deviation of a domain from the symmetric equilateral structure*)


Rot={{Cos[\[Theta]],-Sin[\[Theta]]},{Sin[\[Theta]],Cos[\[Theta]]}};
Rot1={{Cos[\[Phi]1],-Sin[\[Phi]1]},{Sin[\[Phi]1],Cos[\[Phi]1]}};
Rot2={{Cos[\[Phi]2],-Sin[\[Phi]2]},{Sin[\[Phi]2],Cos[\[Phi]2]}};

Loc1={{PtDef1x-size/2},{PtDef1y-size/2}};     (*Note which rotation operation i.e. Rot, Rot1 or Rot2 are in action*)
Loc1R=1*Rot.Loc1;
Loc2={{PtDef2x-size/2},{PtDef2y-size/2}};
```

```
Loc2R=1.Rot.Loc2;
Loc3={{PtDef3x-size/2},{PtDef3y-size/2}};
Loc3R=Rot.Loc3;
(*Loc4={{PtDef4x-size/2},{PtDef4y-size/2}};
Loc4R=Rot2.Rot.Loc4;
Loc5={{PtDef5x-size/2},{PtDef5y-size/2}};
Loc5R=Rot2.Rot.Loc5;
Loc6={{PtDef6x-size/2},{PtDef6y-size/2}};
Loc6R=Rot2.Rot.Loc6;*)

(*prob=5*0.3/(21*4.2);*) "Actually 0.03 gives 21
    triplets for 700x700 Lattice"
(*prob=0.016*)
"Number of triplets"
(*count=Round[prob*size]*)
count=1
"Location of triplets"
(*locations=RandomChoice[Table[i,{i,size}],{count
    ,2}]*)
locations={{size/2,size/2}}

(* Writing Defects: 2B: This is a continuous
    deformation of the magnetic texture at 3
    different locations*)
winding1=1;
TopDef1xR=Round[Loc1R[[1]]]+locations[[1,1]]+0.002;
TopDef1yR=Round[Loc1R[[2]]]+locations[[1,2]]+0.002;

winding2=1;
(*TopDef2xR=Round[Loc2R[[1]]]+locations
    [[1,1]]+0.002;
TopDef2yR=Round[Loc2R[[2]]]+locations[[1,2]]+0.002;
    *)

TopDef2xR=locations[[1,1]]+0.002;
TopDef2yR=locations[[1,2]]+0.002;

(*TopDef2xR=-1*Round[Loc2R[[1]]]+locations
    [[1,1]]+0.002;
TopDef2yR=-1*Round[Loc2R[[2]]]+locations
    [[1,2]]+0.002;*)

winding3=1;
TopDef3xR=Round[Loc3R[[1]]]+locations[[1,1]]+0.002;
TopDef3yR=Round[Loc3R[[2]]]+locations[[1,2]]+0.002;

(*If[DefMany\[Equal]1,For[i=1,i\[LessEqual] size,i
    ++,
For[j=1,j\[LessEqual] size,j++,
LatticeU[[i,j]]=(Exp[\[ImaginaryI]* winding1*ArcTan
    [(i-TopDef1xR),(j-TopDef1yR)]] +Exp[\[ImaginaryI
    ]* winding2*ArcTan[(i-TopDef2xR),(j-TopDef2yR)]]
    + Exp[\[ImaginaryI]* winding3*ArcTan[(i-TopDef3xR
```

```
),(j-TopDef3yR)]])*Exp[-\[ImaginaryI]*(2*\[Pi])
/(1*c)*Mod[(i-size/2)+1*(j-size/2),1*c]]]],Print
["yeehaw"]];
*)

If[DefMany==1,For[i=1,i<= size,i++,
For[j=1,j<= size,j++,
LatticeU[[i,j]]=(Exp[I* winding1*ArcTan[(i-TopDef1xR
),(j-TopDef1yR)]+I* winding2*ArcTan[(i-TopDef2xR)
,(j-TopDef2yR)]+I* winding3*ArcTan[(i-TopDef3xR)
,(j-TopDef3yR)]])*Exp[-I*(2*\[Pi])/(1*c)*Mod[(i-
size/2)+1*(j-size/2),1*c]]]],Print["yeehaw"]];



(*
--------------------------------------------------------------------------
*)
mask=0*IdentityMatrix[size];(*Creating the mask
function makes it possible to define a domain
arrangement of the order defined in LatticeU*)
If [FOV==0,

Monitor[
For[counter=1,counter<(count+1),counter++,
     PtDef1xR=Round[Loc1R[[1]]]+locations[[counter
        ,1]];
     PtDef1yR=Round[Loc1R[[2]]]+locations[[counter
        ,2]];
     PtDef2xR=Round[Loc2R[[1]]]+locations[[counter
        ,1]]+1*1;
     PtDef2yR=Round[Loc2R[[2]]]+locations[[counter
        ,2]]+1*0;
     PtDef3xR=Round[Loc3R[[1]]]+locations[[counter
        ,1]];
     PtDef3yR=Round[Loc3R[[2]]]+locations[[counter
        ,2]];


       (* use this for 3 domain pattern as we had
          been doing, with 5*\[Pi]/180 = \[Theta] and
          Amp1=1;*)(*
       factor=2;
     Subscript[\[Sigma]1, x]=factor/2\[Sqrt](LatCoLen
        /0.6);Subscript[\[Sigma]1, y]=factor/2\[Sqrt
        ](LatCoLen/4);  Subscript[\[Sigma]2, x]=
        factor/2\[Sqrt](LatCoLen/0.6);Subscript[\[
        Sigma]2, y]=factor/2\[Sqrt](LatCoLen/4);
        Subscript[\[Sigma]3, x]=factor/2\[Sqrt](
        LatCoLen/0.6);Subscript[\[Sigma]3, y]=factor
        /2\[Sqrt](LatCoLen/4);*)

```

```
    (*us this for two circular spot with Amp1=0 and \[Theta]=0*)
    Subscript[\[Sigma]1, x]=30;Subscript[\[Sigma]1, y]=30;Subscript[\[Sigma]2, x]=20;Subscript[\[Sigma]2, y]=20;Subscript[\[Sigma]3, x]=20;Subscript[\[Sigma]3, y]=20;

    twist1=-30*\[Pi]/180;Amp1=0;
    a1=(Cos[twist1])^2/(2*Subscript[\[Sigma]1, x]^2)+(Sin[twist1])^2/(2*Subscript[\[Sigma]1, y]^2);
    b1=-Sin[2*twist1]/(4*Subscript[\[Sigma]1, x]^2)+Sin[2*twist1]/(4*Subscript[\[Sigma]1, y]^2);
    c1=(Sin[twist1])^2/(2*Subscript[\[Sigma]1, x]^2)+(Cos[twist1])^2/(2*Subscript[\[Sigma]1, y]^2);
    twist2=-30*\[Pi]/180;Amp2=1;
    a2=(Cos[twist2])^2/(2*Subscript[\[Sigma]2, x]^2)+(Sin[twist2])^2/(2*Subscript[\[Sigma]2, y]^2);
    b2=-Sin[2*twist2]/(4*Subscript[\[Sigma]2, x]^2)+Sin[2*twist2]/(4*Subscript[\[Sigma]2, y]^2);
    c2=(Sin[twist2])^2/(2*Subscript[\[Sigma]2, x]^2)+(Cos[twist2])^2/(2*Subscript[\[Sigma]2, y]^2);
    twist3=-30*\[Pi]/180;Amp3=1;
    a3=(Cos[twist3])^2/(2*Subscript[\[Sigma]3, x]^2)+(Sin[twist3])^2/(2*Subscript[\[Sigma]3, y]^2);
    b3=-Sin[2*twist3]/(4*Subscript[\[Sigma]3, x]^2)+Sin[2*twist3]/(4*Subscript[\[Sigma]3, y]^2);
    c3=(Sin[twist3])^2/(2*Subscript[\[Sigma]3, x]^2)+(Cos[twist3])^2/(2*Subscript[\[Sigma]3, y]^2);

For[i=0,i<size,i++,
For[j=0,j<size,j++,
mask[[i+1,j+1]]=mask[[i+1,j+1]]+
Amp1*Exp[-a1*(i+1-PtDef1xR)^2]*Exp[-2*b1*(i+1-PtDef1xR)*(j+1-PtDef1yR)]*Exp[-c1*(j+1-PtDef1yR)^2]+
Amp2*Exp[-a2*(i+1-PtDef2xR)^2]*Exp[-2*b2*(i+1-PtDef2xR)*(j+1-PtDef2yR)]*Exp[-c2*(j+1-PtDef2yR)^2]+
Amp3*Exp[-a3*(i+1-PtDef3xR)^2]*Exp[-2*b3*(i+1-PtDef3xR)*(j+1-PtDef3yR)]*Exp[-c3*(j+1-PtDef3yR)^2]]];
],Row[{ProgressIndicator[counter,{1,(count+1)}]}," "]]


, Print["Not 3 domains/beamspot model"]]

```

```
If[FOV==1,For[i=0,i<size,i++,
  For[j=0,j<size,j++,
   If[mask[[i+1,j+1]]==0,
    mask[[i+1,j+1]]=1,mask[[i+1,j+1]]=mask[[i+1,j+1]]
   ]
  ]],Print["Not Full Field of View"]]

bSpot=30;
bSpotLoc={{size/2+00,size/2+00}};
If[FOV==2,For[i=0,i<size,i++,
  For[j=0,j<size,j++,
   mask[[i+1,j+1]]=mask[[i+1,j+1]]+Exp[-(1/(2*(bSpot)^2)) (i+1 - bSpotLoc[[1,1]])^2]*Exp[-(1/(2*(bSpot)^2)) (j+1 - bSpotLoc[[1,2]])^2];
  ]],Print["Not a short field of view"]]

(*
Calculate Domain area from \[Pi]*a*b formula and cross verify it with counting pixels. Then with this we can absolutely define the size of the blots
*)


(*
--------------------------------------------------------------------------------------*)
(*"Creating random un-corordinated domains of such count at the following locations"
countrandom=Round[count*1/count]
locations2=RandomChoice[Table[i,{i,size}],{countrandom,2}]
Monitor[
For[counter=1,counter<(countrandom+1),counter++,
    PtDef1xR=Round[Loc1R[[1]]]+locations2[[counter,1]];
    PtDef1yR=Round[Loc1R[[2]]]+locations2[[counter,2]]+\[Delta];

      fact=1.45;
    Subscript[\[Sigma]4, x]=fact/2\[Sqrt](LatCoLen/0.7);Subscript[\[Sigma]4, y]=fact/2\[Sqrt](LatCoLen/4);

    twist1=-30*\[Pi]/180;Amp1=1;
    a1=(Cos[twist1])^2/(2*Subscript[\[Sigma]4, x]^2)+(Sin[twist1])^2/(2*Subscript[\[Sigma]4, y]^2);
    b1=-Sin[2*twist1]/(4*Subscript[\[Sigma]4, x]^2)+Sin[2*twist1]/(4*Subscript[\[Sigma]4, y]^2);
```

```
    c1=(Sin[twist1])^2/(2*Subscript[\[Sigma]4, x]^2)+(Cos[twist1])^2/(2*Subscript[\[Sigma]4, y]^2);
For[i=0,i<size,i++,
  For[j=0,j<size,j++,
  mask[[i+1,j+1]]=mask[[i+1,j+1]]+(Amp1*Exp[-a1*(i+1-PtDef1xR)^2]*Exp[-2*b1*(i+1-PtDef1xR)*(j+1-PtDef1yR)]*Exp[-c1*(j+1-PtDef1yR)^2])]];
],Row[{ProgressIndicator[counter,{1,(countrandom+1)}]}," "]]*)

(*For[i=0,i<size,i++,
  For[j=0,j<size,j++,
   If[mask[[i+1,j+1]]\[Equal]0,
    mask[[i+1,j+1]]=1,mask[[i+1,j+1]]=mask[[i+1,j+1]]
   ]
  ]]*)
(*
   ----------------------------------------------------------------------------------------------------
   *)

(* Cutting off overlapped filters to a maximum of 1 *)
For[i=0,i<size,i++,
  For[j=0,j<size,j++,
   If[mask[[i+1,j+1]]>1,
    mask[[i+1,j+1]]=1,mask[[i+1,j+1]]=mask[[i+1,j+1]]
   ]
  ]]
For[i=0,i<size,i++,
For[j=0,j<size,j++,
    If[mask[[i+1,j+1]]<1*10^-4,
  mask[[i+1,j+1]]=0,mask[[i+1,j+1]]=mask[[i+1,j+1]]
    ]
]]

"How the mask looks"
MASK=Rotate[Image[mask],\[Pi]/2]

ListVectorPlot[Table[{{m,n},{Re[LatticeU[[m,n]]],Im[LatticeU[[m,n]]]}},{m,Round[size/2]-10,Round[size/2]+10,1},{n,Round[size/2]-10,Round[size/2]+10,1}],VectorPoints->All,VectorMarkers->"Arrow",VectorScale->0.02]

If[Jitter==1,
For[i=1,i<= size,i++,
For[j=1,j<= size,j++,
LatticeU[[i,j]]=LatticeU[[i,j]]*Exp[-I*Angl*(0.5-RandomReal[])]]];
,Print["No Jitter"]]
```

```
"After Randomness"
ListVectorPlot[Table[{{m,n},{Re[LatticeU[[m,n]]],Im[
    LatticeU[[m,n]]]}},{m,Round[size/2]-10,Round[size
    /2]+10,1},{n,Round[size/2]-10,Round[size
    /2]+10,1}],VectorPoints->All,VectorMarkers->"
    Arrow",VectorScale->0.02]

(*ImageCrop[Image[mask],{210,210}]*)

For[i=1,i<= size,i++,
For[j=1,j<= size,j++,
Lattice[[i,j]]=LatticeU[[i,j]]*mask[[i,j]]]]; (*
    Creating the desired domain arrangement as the
    mask function allows*)


If[Defect==1,
For[i=1,i<= size,i++,
For[j=1,j<= size,j++,
If[RandomReal[]<0.1,
Lattice[[i,j]]=0*Lattice[[i,j]],Lattice[[i,j]]=1*
    Lattice[[i,j]];
     ];
     ]];

,Print["No Random Defect"]];



If[c==1,FM=1,FM=0]

(*
    -----------------------------------------------------------------
    *)
Lattice[[Round[size/2],Round[size/2]]];
(*"The Domain arrangement"
ListVectorPlot[Table[{{m,n},{Re[Lattice[[m,n]]],Im[
    Lattice[[m,n]]]}},{m,1,size,1},{n,1,size,1}],
    VectorPoints\[Rule]All,VectorMarkers\[Rule]"Arrow
    ",VectorScale\[Rule]0.05]
*)

"Note LatticeVectorPlot flips the image on x axis
    comapared to the picture generated by Image"
ReciLattice=(Abs[Fourier[Lattice*(-1)^Table[FM*(i+j)
    ,{i,size},{j,size}]]])^2;
RIM=Image[ReciLattice]
RIM1=Image[ReciLattice/Max[ReciLattice]]
RIM2=Rotate[RIM,\[Pi]/2];
```

Appendix B
Placing the Detector Image in Reciprocal Space

This code places a detector image and places it in the reciprocal space (k-space) after learning the diffraction geometry. User need to put the experimental details and the sample details as inputs. Any line cut can be taken to evaluate the modulation separation for the dilute domain scenario which gives us the linear dimensions of real space separation between domains. Finally it places the detector image in the H-K-L space. One can take projection plane after this.

```
    close all;
clear all;
[inputfile,inputpath]=uigetfile('*.jpg');%Enter the
    file format here if user wants to change
inputpath=strcat(inputpath,inputfile);
image=imread(inputpath);


figure('Name','Input Image');imshow(image);cmap=
    colormap;%Converting to grayscale for now
figure('Name','Grayscaled Image');imageGray=
    im2double(rgb2gray(image));imshow(imageGray);
[row,col]=size(imageGray);

a=3.86;b=3.86;c=3.86;alpha=90*pi/180;beta=90*pi/180;
    gamma=90*pi/180;%Unit vectors for Pseudocubic PNO
     on LSAT

c_vec=c*[0;0;1];
b_vec=b*cos(alpha)*[0;0;1]+b*sin(alpha)*[0;1;0];
a_vec=a*cos(beta)*[0;0;1]+a*((cos(gamma)-cos(beta)*
    cos(alpha))/sin(alpha))*[0;1;0]+a*(sqrt(1-(cos(
    gamma))^2-(cos(beta))^2-(cos(alpha))^2+2*cos(
    gamma)*cos(beta)*cos(alpha))/sin(alpha))*[1;0;0];

Volume=transpose(a_vec)*cross(b_vec,c_vec);

astar_vec=(2*pi/Volume)*cross(b_vec,c_vec);
```


R. Basak, *Nucleation, Fabrication, and Competition in Antiferromagnetic Textures Studied with Coherent Resonant Soft X-ray Scattering*, Springer Theses,
https://doi.org/10.1007/978-3-032-19199-1

```
bstar_vec=(2*pi/Volume)*cross(c_vec,a_vec);
cstar_vec=(2*pi/Volume)*cross(a_vec,b_vec);

astar=norm(astar_vec);bstar=norm(bstar_vec);cstar=norm(cstar_vec);
alphastar=acos((cos(beta)*cos(gamma)-cos(alpha))/(sin(beta)*sin(gamma)));
betastar=acos((cos(gamma)*cos(alpha)-cos(beta))/(sin(gamma)*sin(alpha)));
gammastar=acos((cos(alpha)*cos(beta)-cos(beta))/(sin(gamma)*sin(alpha)));


maxValue = max(imageGray(:));
[rowsOfMaxes,colsOfMaxes] = find(imageGray == maxValue);hand=size(rowsOfMaxes);
figure('Name','Location of Maxes');
imshow(imageGray);hold on; scatter(colsOfMaxes(:),rowsOfMaxes(:),'*r');
for count=1:1:hand(1,1)
    txt_X=num2str(colsOfMaxes(count,1));
    txt_Y=strcat(',',num2str(rowsOfMaxes(count,1)));
    txt=strcat(txt_X,txt_Y);
    text(colsOfMaxes(count,1),rowsOfMaxes(count,1),txt);
end
title('Maximum values of the pixels observed at following (col,row) pixels');

%%%% Prompter to get experimental details from user
prompt={'Peak Y (row)','Peak X (col)','2Theta (in Deg):','Chi (in Deg): 90||Normal','Phi (in Deg): from x','Energy (in eV):','pixel width(in m):','pixel height(in m):','Detector distance from sample (in m):'};
input=inputdlg(prompt);
[rowip,colip]=size(input);
for count=1:1:rowip
    input{count}=str2double(input{count});
end


TwoTheta=zeros(row,1);Chi=zeros(col,1);%Initialize
Y_center=input{1};
X_center=input{2};
two_theta_peak=input{3}*(pi)/180;
chi_peak=input{4}*pi/180;
phi_peak=input{5}*(pi)/180;
Lambda=12380/input{6};
lW=input{7};
lH=input{8};
r_dist=input{9};%Assign
```

```
k=2*(pi)/Lambda;%evaluate k magnitude from Lambda calculated from Energy



for Y=1:1:row%Assign every pixel a Two Theta value
    TwoTheta(Y,1)=two_theta_peak+(Y-Y_center)*lH/r_dist;
end
for X=1:1:col% Assign every pixel a Chi Rotation value
     Chi(X,1)=0 +(X-X_center)*lW/r_dist;
end


[Ch,TwoTh]=meshgrid(Chi,TwoTheta);%This assigns Chi_Rot and TwoTheta values to every pixel of the image
% figure('Name','Detector Image in radian');
% fig=surf(Chi,TwoTh,imageGray,'EdgeColor','none');
% colormap parula;xlabel('Phi (in rad)');ylabel('2 Theta (in rad)');title('Detector image in angle space');

%% Now our task is to create this image on the relevant plane on  kx-ky-kz coordinate system.
% Dictionary of translation, Chi value is angular Chi deviation from 90 deg
% measured from +kz, since it does not matter when placing the image in
% kspace (for k-space is isotropic) all we care is magnitude, I think.
% Theta from kx-ky plane, TO TAKE A PROJECTION replace k for that axis with 0

kx=k*(sin(TwoTh/2).*sin(Ch));% What we have used so far
ky=k*(cos(TwoTh/2)-cos(two_theta_peak/2));
kz=k*(sin(TwoTh/2).*cos(Ch)+sin(two_theta_peak/2));

% kx=k*(sin(Ch));% What Makes most sense
% ky=k*(cos(TwoTh/2).*cos(Ch)-cos(two_theta_peak/2));
% kz=k*(sin(TwoTh/2).*cos(Ch)+sin(two_theta_peak/2));

% kx=k*(sin(Ch).*sin(TwoTh/2));% What the NatPhotonic paper uses
% ky=k*(cos(TwoTh/2).*cos(Ch)-cos(two_theta_peak/2));
% kz=k*(sin(TwoTh/2)+sin(two_theta_peak/2));

```

```
% kx=k*((cos(Th).*cos(Ph))-(cos(theta_peak).*cos(
    phi_peak)));
% ky=k*((cos(Th).*sin(Ph))-(cos(theta_peak).*sin(
    phi_peak)));
% kz=k*(sin(Th)+sin(theta_peak));

%The k-space image is not oriented in HKL space but
    it gives the pixels a
%reciprocal space metric so that there is a
    meaningful separation vector in
%reciprocal space between two pixels. To orient our
    2D image properly we
%need to know our alignment. That comes next.
figure('Name','Detector Image in k-space');
surf(kx,ky,kz,imageGray,'EdgeColor','none');
caxis([min(min(imageGray)), max(max(imageGray))]);
    colormap(cmap);colorbar;xlabel('kx');ylabel('ky')
    ;zlabel('kz');title('Detector image in k-space');

%%%% Taking cuts and correlation lengths.
%%%% Here we will take cuts and measure correlation
    length corresponding to
%%%% inverse of the
figure();figu=imagesc(Chi,TwoTheta/2,imageGray);
    colorbar;xlabel('Chi in rad');ylabel('Theta in
    rad');
% title('Select 2 points & hit "Enter" to make a cut
    ')
% profileChi=[0.0062 -0.0063];
% profileTheta=[0.9531 0.9668];
%[cutChi,cutTheta,cutIntensity]=improfile(figu.XData
    ,figu.YData,figu.CData,[0.0172 -0.0136],[0.9433
    0.9745],709);
%EightyK
%[cutChi,cutTheta,cutIntensity]=improfile(figu.XData
    ,figu.YData,figu.CData
    ,[0.0062:-0.000025*2:-0.0063],[0.9531:2.74e
    -5*2:0.9668]);%For
%Hex
% [cutChi,cutTheta,cutIntensity]=improfile(figu.
    XData,figu.YData,figu.CData,[0.0059
    -0.0077],[0.9525 0.9674],380);
%wavey
[cutChi,cutTheta,cutIntensity]=improfile
hold on;
scatter(cutChi(:),cutTheta(:),'.r');

cutkx=k*(cos(cutTheta).*sin(cutChi));%what we have
    been using
cutky=k*(cos(cutTheta)-cos(two_theta_peak/2));
cutkz=k*(sin(cutTheta).*cos(cutChi)+sin(
    two_theta_peak/2));

```

```
% cutkx=k*sin(cutChi);%What makes most sense
% cutky=k*(cos(cutTheta).*cos(cutChi)-cos(two_theta_peak/2));
% cutkz=k*(sin(cutTheta).*cos(cutChi)+sin(two_theta_peak/2));

% cutkx=k*(sin(cutChi).*sin(cutTheta));%What the NatPhotonics paper uses
% cutky=k*(cos(cutTheta).*cos(cutChi)-cos(two_theta_peak/2));
% cutkz=k*(sin(cutTheta)+sin(two_theta_peak/2));

%cutkx=k*((cos(cutTheta).*cos(cutChi))-(cos(theta_peak).*cos(phi_peak)));
%cutky=k*((cos(cutTheta).*sin(cutChi))-(cos(theta_peak).*sin(phi_peak)));
%cutkz=k*(sin(cutTheta)+sin(theta_peak));

kmag=sqrt((cutkx.*cutkx)+(cutky.*cutky)+1*(cutkz.*cutkz));
figure();
plot(kmag,cutIntensity);
xlabel('distance in k-space along the cut ($\AA^{-1}$)','Interpreter','latex');ylabel('Intensity');title('Cuts Cuts Cuts');
%%%% Add a figure which shows where we are taking the cut in H-K-L space
%%%% and also k space
%% q_HKL = 2*pi/d_HKL; d_HKL=a/(H^2+K^2+L^2) = 3.86/(0.25*sqrt(3)) = 8.914 --> 0.705 (Ang^-1)
%% Orienting the 2D image in HKL space

% Let's make a transformation matrix B which translates a description of a
% vector in reciprocal lattice vectors to the description of that vector in
% terms of crystal or sample cartesian system.This is the end result here.
B=[astar bstar*cos(gammastar) cstar*cos(betastar);0 bstar*sin(gammastar) -cstar*sin(betastar)*cos(alpha);0 0 (2*pi)/c];

Reci_metric_tensor=[astar_vec.'*astar_vec astar_vec.'*bstar_vec astar_vec.'*cstar_vec;bstar_vec.'*astar_vec bstar_vec.'*bstar_vec bstar_vec.'*cstar_vec;cstar_vec.'*astar_vec cstar_vec.'*bstar_vec cstar_vec.'*cstar_vec];

% %It is interesting that alternatively that can also be found by
% Reci_metric_tensor_alt=transpose(B)*B;
% Reci_metric_tensor is used to define any dot product in Reciprocal Space
```

```
Magn=@(t) sqrt([t(1,1) t(2,1) t(3,1)]*Reci_metric_tensor*[t(1,1);t(2,1);t(3,1)]);% Magnitude function for RLVs
Angl=@(a,b) acos(([a(1,1) a(2,1) a(3,1)]*Reci_metric_tensor*[b(1,1);b(2,1);b(3,1)])/(Magn(a)*Magn(b)));% Angle Calculator b2in 2 RLVs

% or0 and or1 decides the scattering plane, i.e. the plane accessed by th
% and 2th motors when chi=90 and phi=0. or2 is the direction of chi motor
or0=[1;1;1];
or1=[1;-1;0];%Th Direction
or2=[1;1;-2];%Chi Direction

for X=1:1:col% Assign every pixel a Chi_Rotation value
    Chi(X,1)=0+(X-X_center)*1W/r_dist;
end

% or0=[0;0;1];%To evluate from the experimental starting condition
% or1=[1;0;0];
% or2=[0;1;0];
%
% for X=1:1:col% Assign every pixel a Chi_Rotation value
%     Chi(X,1)=((pi/2)-chi_peak)+(X-X_center)*1W/r_dist;
% end

unit_or0=(1/Magn(or0))*or0;
unit_or1=(1/Magn(or1))*or1;
unit_or2=(1/Magn(or2))*or2;

U_theory=[[1;0;0] [0;1;0] [0;0;1]]/[(B*unit_or0) (B*unit_or1) (B*unit_or2)];% Check U'*U =Identity(3)
UB_theory=U_theory*B;

CHI = @(t) [cos(t) 0 sin(t) ; 0 1 0 ; -sin(t) 0 cos(t)] ;
OMEGA = @(t) [cos(t) -sin(t) 0 ; sin(t) cos(t) 0 ; 0 0 1] ;
PHI=@(t) [1 0 0; 0 cos(t) -sin(t);0 sin(t) cos(t)];% unused

[Ch,TwoTh]=meshgrid(Chi,TwoTheta);

H=zeros(row,col);K=zeros(row,col);L=zeros(row,col);
%If this is a q-Chi scan
% for i=1:1:row
```

```
%       for j=1:1:col
%           vecto=(eye(3)/(OMEGA(0)*CHI(Ch(i,j))*PHI(0)*UB_theory))*([(4*pi/Lambda)*(sin(0.5*TwoTh(i,j)));0;0]);
%           H(i,j)=vecto(1,1);
%           K(i,j)=vecto(2,1);
%           L(i,j)=vecto(3,1);
%       end
% end

% If this was a Th-Chi distribution
% for i=1:1:row
%       for j=1:1:col
%           vecto=(eye(3)/(OMEGA(0.5*TwoTh(i,j)-two_theta_peak/2)*CHI(Ch(i,j))*PHI(0)*UB_theory))*([(4*pi/Lambda)*(sin(two_theta_peak/2));0;0]);
%           H(i,j)=vecto(1,1);
%           K(i,j)=vecto(2,1);
%           L(i,j)=vecto(3,1);
%       end
% end

% If this is a 2th-chi distribution
for i=1:1:row
    for j=1:1:col
        vecto=(eye(3)/(OMEGA(-0.5*TwoTh(i,j)+two_theta_peak/2)*CHI(Ch(i,j))*PHI(0)*UB_theory))*([(4*pi/Lambda)*(sin(0.5*TwoTh(i,j)));0;0]);
        H(i,j)=vecto(1,1);
        K(i,j)=vecto(2,1);
        L(i,j)=vecto(3,1);
    end
end
%

figure();
surf(H,K,L,1*imageGray,'EdgeColor','none');
colormap inferno;
hold on
%quiver3(0.24,0.24,0.24,0.01325,0.01325,0.01325,'LineWidth',2);%q Vector Direction!
xlabel('H');ylabel('K');zlabel('L');
edge1=[H(row,1)-H(1,1);K(row,1)-K(1,1);L(row,1)-L(1,1)];%Detector Edge 1
quiver3(H(1,1),K(1,1),L(1,1),edge1(1,1),edge1(2,1),edge1(3,1),0,'LineWidth',2,'Color',[1 0.9 0]);
edge2=[H(1,col)-H(1,1);K(1,col)-K(1,1);L(1,col)-L(1,1)];%Detector Edge 2
quiver3(H(1,1),K(1,1),L(1,1),edge2(1,1),edge2(2,1),edge2(3,1),0,'LineWidth',2,'Color',[1 0.5 0]);
quiver3(0.25,0.25,0.24,0,0,0.0225,'LineWidth',2);%Sample Growth direction
```

```
detNorm=cross(edge1,edge2);
detNorm=detNorm/norm(detNorm);%unit_kOUT
quiver3(0.2505,0.2342,0.2423,0.022*detNorm(1,1),0.022*detNorm(2,1),0.022*detNorm(3,1),'LineWidth',2);

title('Detector in HKL Space');
text(0.2425,0.245,0.245,'Magnetic Order Q');
text(0.25,0.25,0.2625,'Sample Normal');
text(0.2505,0.24,0.245,'kOUT');
str1=num2str(round(edge1/min(edge1),1));%Change min - max as required
str2=num2str(round(edge2/max(edge2),1));
text(H(row,1)+0.0025,K(row,1),L(row,1),str1);
text(H(1,col)-0.0025,K(1,col),L(1,col),str2);
axis([0.235 0.265 0.235 0.265 0.235 0.265])
pbaspect([1 1 1])% view([135 45])%view from Magnetic Order Q
view([0 90])% view from the Side of the Detector
```

The manufacturer's authorised representative in the EU is Springer Nature Customer Service Centre GmbH, Europaplatz 3, 69115 Heidelberg, Germany. If you have any concerns regarding our products, please contact ProductSafety@springernature.com

Printed and bound by CPI Group (UK) Ltd, Croydon, CR0 4YY
17/09/2026
02218437-0001